MICROSCOPY HANDBOOKS 36

Lectin Histochemistry

Royal Microscopical Society MICROSCOPY HANDBOOKS

Lectin Histochemistry

a concise practical handbook

S.A. Brooks
Oxford Brookes University,
Oxford, UK

A.J.C. Leathem
University College London Medical School,
London, UK

U. Schumacher
University of Southampton,
Southampton, UK

In association with the Royal Microscopical Society

© BIOS Scientific Publishers Limited, 1997

First published 1997

A CIP catalogue record for this book is available from the British Library.

ISBN 1 85996 100 2

BIOS Scientific Publishers Ltd,
9 Newtec Place, Magdalen Road, Oxford OX4 1RE, UK.
Tel: +44 (0) 1865 726286, Fax: +44 (0) 1865 246823.
World Wide Web home page: http://www.Bookshop.co.uk/BIOS/

This book is dedicated to our parents; Kenneth and Miriam Brooks, Bill and Barbara Leathem and Robert and Hertha Schumacher.

DISTRIBUTORS

Australia and New Zealand
    DA Information Services
    648 Whitehorse Road, Mitcham
    Victoria 3132

India
    Viva Books Private Ltd
    4325/3 Ansari Road
    Daryaganj
    New Delhi 110002

Singapore and South East Asia
    Toppan Company (S) PTE Ltd
    38 Liu Fang Road, Jurong
    Singapore 2262

USA and Canada
    BIOS Scientific Publishers
    PO Box 605,
    Herndon, VA 20172–0605

Typeset by Chandos Electronic Publishing, Stanton Harcourt, Oxon, UK.
Printed by Information Press, Eynsham, Oxon, UK.

Contents

Abbreviations

Lectins

Below are given the abbreviations used in the text. A more comprehensive list of lectins, including abbreviations and details of sugar binding characteristics is given in Table 2.1.

Abbreviation		Common name	Latin name
APA	*Abrus precatorius* agglutinin	jequirity bean	*Abrus precatorius*
BPA	*Bauhinia purpurea* agglutinin	camel's foot tree	*Bauhinia purpurea*
BSA	*Bandeiraea simplicifolia* agglutinin	–	*Bandeiraea simplicifolia* or *Griffonia simplicifolia*
BSA-I	*Bandeiraea simplicifolia* agglutinin isolectin I		
Con A	Concanavalin A	jack bean	*Canavalia ensiformis*
DBA	*Dolichos biflorus* agglutinin	horse gram	*Dolichos biflorus*
DSA	*Datura stramonium* agglutinin	jimson weed or thorn apple	*Datura stramonium*
ECA	*Erythrina cristagalli* agglutinin	coral tree	*Erythrina cristagalli*
GNA	*Galanthus nivalis* agglutinin	snowdrop	*Galanthus nivalis*

Abbreviation		Common name	Latin name
HAA	*Helix aspersa* agglutinin	garden snail	*Helix aspersa*
HPA	*Helix pomatia* agglutinin	Roman or edible snail	*Helix pomatia*
IAA	*Iberis amara* agglutinin	–	*Iberis amara*
LEA	*Lycopersicon esculentum* agglutinin	tomato	*Lycopersicon esculentum*
LCA	*Lens culinaris* agglutinin	lentil	*Lens culinaris*
LPA	*Limulus polyphemus* agglutinin	horseshoe crab	*Limulus polyphemus*
LTA	*Lotus tetragonolobus* agglutinin	asparagus pea	*Lotus tetragonolobus*
MAA	*Maackia amurensis* agglutinn	maackia	*Maackia amurensis*
MPA	*Maclura pomifera* agglutinin	osage orange	*Maclura pomifera*
PHA	phytohaem-agglutinin	red kidney bean	*Phaseolus vulgaris*
PHA-L	phytohaem-aggutinin-leucoagglutinin		
PHA-E	phytohaem-agglutinin-erthyrocyte agglutinin		
PNA	peanut agglutinin	peanut	*Arachis hypogaea*

| PSA | *Pisum sativum* agglutinin | garden pea | *Pisum sativum* |

Abbreviation		*Common name*	*Latin name*
PWA	pokeweed agglutinin	pokeweed	*Phytolacca americana*
PWM	pokeweed mitogen		
RCA	*Ricinus communis* agglutinin	castor oil bean	*Ricinus communis*
SBA	soyabean agglutinin	soyabean	*Glycine max*
SNA	*Sambucus nigra* agglutinin	elderberry	*Sambucus nigra*
WGA	wheatgerm agglutinin	wheatgerm	*Triticum vulgare*
UEA	*Ulex europaeus* agglutinin	gorse or furze	*Ulex europaeus*
UEA-I	*Ulex europaeus* agglutinin isolectin I		
VAA	*Viscum album* agglutinin	European mistletoe	*Viscum album*
VVA	*Vicia villosa* agglutinin	hairy vetch	*Vicia villosa*

Monosaccharides

fuc	fucose
gal	galactose
galNAc	*N*-acetyl galactosamine
glc	glucose
glcNAc	*N*-acetyl glucosamine
man	mannose
NANA	*N*-acetyl neuraminic acid (sialic acid)
neuAc	neuraminic acid (sialic acid)

Chemicals

AEC	amino ethyl carbozole
APS	ammonium persulphate
BSA	bovine serum albumin
DAB	diaminobenzidine
DAPI	4,6-diamino-2-phenylindole
DMF	dimethylformamide
DPX	dibutyl polystyrene xylene
EDTA	ethylenediaminetetracetic acid
FITC	fluorescein isothiocyanate
FLUOS	5(6)-carboxyfluorescein-N-hydroxysuccinomide ester
NBT	nitro blue tetrazolium
OCT	'optimum cutting temperature' sectioning medium
TCA	trichloroacetic acid
TEMED	N,N,N',N'-tetramethylethylenediamine
TRITC	tetramethylrhodamine isothiocyanate

General

ABC	avidin biotin complex
APAAP	alkaline phosphatase anti-alkaline phosphatase
CBB	Coomassie Brilliant Blue
CFA	complete Freud's adjuvant
CIN	cervical intraepithelial neoplasia
DTIF	direct tissue isoelectric focusing
EM	electron microscopy
GAG	glycosaminoglycans
IFA	incomplete Freud's adjuvant
IEF	isoelectric focusing
PAP	peroxidase anti-peroxidase
PAS	periodic acid–Schiff
PBS	phosphate buffered saline
pI	isoelectric point
SDS–PAGE	sodium dodecyl sulphate polyacrylamide gel electrophoresis
TBS	Tris-buffered saline
v/v	volume per volume
w/v	weight per volume

Preface

Lectin histochemistry is used to localize carbohydrate residues (on glycoproteins, glycolipids and glycosaminoglycans) in cell and tissue preparations. With recent advances in molecular biology, it is becoming increasingly apparent that carbohydrates are of fundamental importance in a whole range of cell recognition events in biological mechanisms as diverse as pollination in plants, fertilization in animals, inflammation, infection and in diseases such as cancer and rheumatoid arthritis. With the current explosion of interest in glycobiology and glycoscience, lectins are essential tools in mapping and understanding carbohydrate expression. They are, however, often poorly understood and inefficiently used.

This succinct and highly practical guide is aimed at laboratory based scientists and students with little or no previous experience of using lectins. Common misconceptions about lectins and the way in which they work are frequently responsible for badly designed experiments, technical errors, and poor results. Key features of this volume are an emphasis on practical detail, hints and tips, troubleshooting sections, and highlighting of the ways in which lectin histochemistry differs from 'traditional' histochemical methods. The reader is taken step by step through each technique, with a wealth of technical and practical detail.

The book covers basic theory regarding lectins and the carbohydrate structures they recognize, what lectins can be used for and why they should be used, lectin histochemistry at light and electron microscope level, analysis of glycoproteins by use of electrophoretic techniques and lectin blotting, analysis and interpretation of lectin binding, applications and usefulness of lectins, and the use of histochemistry to localize endogenous tissue lectins.

The study of carbohydrate expression by cells and tissues is a fascinating and expanding field. It is hoped that this volume will provide essential technical background to facilitate successful and worthwhile investigations.

S.A. Brooks
A.J.C. Leathem
U. Schumacher

1 An Introduction to the Field

1.1 Definition

Lectins are naturally occurring proteins and glycoproteins which selectively bind non-covalently to carbohydrate residues. It is for this reason that they are of interest and use in histochemistry, as they can be employed as exquisitely specific probes to localize defined monosaccharides and oligosaccharides from the astoundingly heterogeneous mixture of carbohydrate residues on or in cells and the extracellular matrix.

The term 'lectin' was first proposed by Boyd and Shapleigh in 1954. It is derived from the Latin verb *'legere'*, which means to pick out, select or choose, and refers to the remarkable selectivity and specificity with which lectins recognize and bind to carbohydrate structures. It has largely superseded historical terms, such as 'phytohaemagglutinin' or 'agglutinin' (originally used to describe the ability of lectins to cross-link and agglutinate cells).

The most widely accepted definition of the term 'lectin' is that originally proposed by Goldstein *et al.* (1980) and adopted by the Nomenclature Committee of the International Union of Biochemistry. This definition states that a lectin is 'a carbohydrate binding protein of non-immune origin, that agglutinates cells and/or precipitates polysaccharides or glycoconjugates'.

This definition implies that lectins are:

- multivalent – two or more sugar binding sites are necessary for the cross-linking of cells in agglutination, or of polysaccharides/glycoconjugates in precipitation. However, in recent years, monovalent carbohydrate-binding molecules such as galectin 3 and adhesins have been described which are monovalent molecules but show multivalent behaviour when they are located close together at the cell surface;

1

- not antibodies – that they are of 'non-immune origin' distinguishes lectins from antibodies directed against carbohydrate antigens, which can also act as agglutinins;
- not enzymes – the definition excludes most sugar binding enzymes (e.g. glycosidases, glycosyl transferases, glycosyl kinases) as they are monovalent, rather than polyvalent. In addition, given the right conditions, enzymes will catalyse a reaction so that sugars are modified – this will not happen when lectins bind to carbohydrates;
- distinct from certain toxins – for example, the toxic A chain of ricin, which possess only one sugar binding site, and acts as an enzyme.

1.2 A brief history of lectinology

Stillmark was the first to describe the activity of a lectin. During work for his doctoral thesis in 1887–1888 at the University of Dorpat in Estonia, one of the oldest universities in Czarist Russia, he investigated extracts of seeds from the *Euphorbiaceae*. He isolated a toxic protein, which he called ricin, from the seeds of the castor oil plant *Ricinus communis*, and tested its effect upon erythrocytes, liver cells, leukocytes and epithelial cells. He noted an agglutination reaction 'like in clotting'. Erythrocytes from different species reacted in different ways. Over the next 40 years or so, a great many papers appeared describing dozens of new lectins (see Kocourek, 1986, for an excellent review). Owing to ease of availability, most were derived from plant sources, but lectins were also discovered subsequently to be present in bacteria, viruses, fungi, the blood and fluids of invertebrates, in the venom of snakes and even in man.

Stillmark, in his thesis of 1888, noted that the presence of serum effectively inhibited haemagglutination by ricin. Other scientists followed up this observation, noting, for example, that gastric mucin had the same inhibitory effect (Landsteiner and Raubitschek, 1909). The significance of these observations were overlooked at the time, but they were the first indication of the carbohydrate-binding nature of lectins. Watkins and Morgan (1952) more than 40 years later were the first to show that monosaccharides were capable of inhibiting lectin activity.

Landsteiner and Raubitschek (1907) were perhaps the first to note that lectins did not always agglutinate red blood cells of different individuals to an equal extent. It was not, however, until 30 years later that the reason for this became apparent with the first report of blood group-specific lectins. A pioneering paper was that of Boyd and

Raguera (1949), which described blood group A-specific haemagglutination by an extract of Lima beans, *Phaseolus lunatus* syn. *limensis*. This led to enormous interest in the search for blood group-specific lectins and between 1945 and 1964 over 100 of them were discovered. *Table 1.1* gives a list of some blood group-specific lectins. The principle by which multivalent lectin binds to cell surface carbohydrates and cross-links cells in an agglutination reaction is illustrated in *Figure 1.1*.

Morgan and Watkins (1959) were the first to demonstrate that blood group specificity of lectins was as a direct result of their sugar binding specificity. They showed that haemagglutination by blood group A-specific lectin from *P. lunatus* syn. *limensis*, the Lima bean, could be inhibited by galNAc, while agglutination by blood group O-specific *Lotus tetragonolobus* lectin (LTA) could be inhibited by fuc. This was one of the first pieces of evidence for the presence of carbohydrates on the cell surface. The human A, B, and O blood group sugars are illustrated in *Figure 1.2*.

1.3 Nomenclature

Lectins are usually named after their source; either by its common name (e.g. peanut lectin) or by its Latin name (e.g. *Tetragonalobus*

Table 1.1: Examples of lectins that bind human blood group sugars

Source of lectin			
Latin name	**Common name**	**Abbreviation**	**Blood group**
Helix pomatia	Roman snail	HPA	A
Helix aspersa	Garden snail	HAA	A
Phaseolus lunatus	Lima bean	LBA	A
Vicia villosa	Hairy vetch	VVA	A
Dolichos biflorus	Horse gram	DBA	A_1
Sophora japonica	Japanese pagoda tree	SJA	B
Bandeiraea simplicifolia I (also called *Griffonia simplicifolia*)	—	BSA-I	B
Tetragonolobus purpureas	Asparagus pea	—	O/H
Ulex europaeus I	Gorse or furze	UEA-I	O/H
Iberis amara	—	IAA	M
Vicia graminea	—	VGA	N
Arachis hypogaea	Peanut	PNA	T
Salvia horminum	Salvia	SHA	T and Cad
Salvia sclarea	Salvia	SSA	Tn and Cad
Bandeiraea simplicifolia II	—	BSA-II	Tk

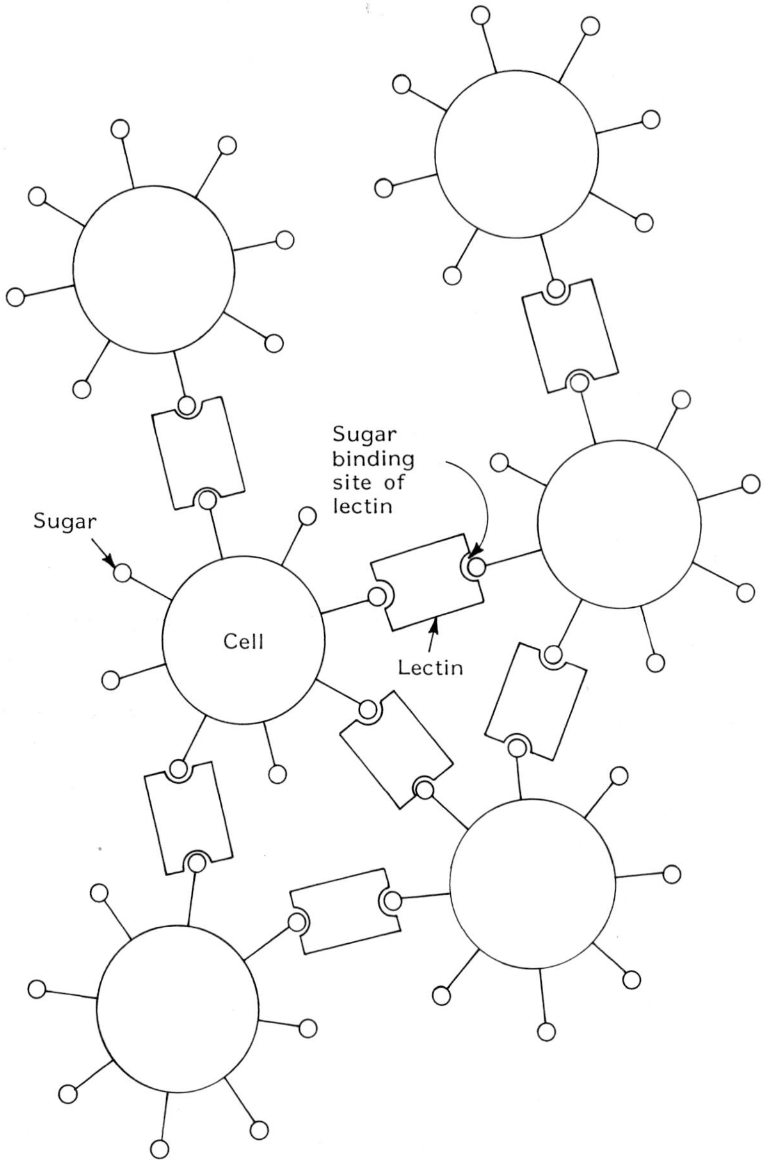

Figure 1.1:   Lectin-mediated cell agglutination.
   In this simple example a divalent lectin binds to cell surface carbohydrates and cross-links the cells in an agglutination reaction.

purpureas lectin). Often these names are abbreviated, for example soyabean lectin is often referred to as SBA (which stands for soya bean agglutinin – agglutinin being a historical term for lectin); the

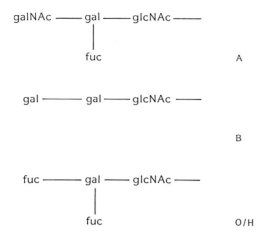

Figure 1.2:   Human A, B and O blood group sugars.

lectin from *Helix pomatia*, the Roman snail, is commonly referred to as HPA (*Helix pomatia* agglutinin). Some abbreviations can be confusing, for example, the lectin from *Phaseolus vulgaris* is often, on historical grounds, quite inappropriately known as PHA, which actually stands for phytohaemmagglutinin. Peumans and Van Damme (1994) have recently proposed a new nomenclature system for lectins which incorporates simple abbreviations for the source of the lectin and its monosaccharide binding specificity. This would appear to be an excellent idea, but has not yet been generally adopted.

1.4 Simple and complex sugars

The chemistry and nomenclature of carbohydrates is a complex subject which lies beyond the scope of this guide (the interested reader is advised to consult any good biochemistry textbook), but a simple description of the different simple and complex sugar types is given here.

1.4.1 Monosaccharides

The sugar chains in human and mammalian tissues are generally made up of combinations of seven simple sugar 'building blocks' called monosaccharides. These are mannose (man), glucose (glc), galactose (gal), fucose (fuc), *N*-acetyl galactosamine (galNAc), *N*-acetyl glucosamine (glcNAc), and sialic or neuraminic acids (NeuAc, neuraminic acid or NANA, *N*-acetyl neuraminic acid). The structure of these monosaccharides is illustrated in *Figure 1.3*. Additional monosaccharides occur in other species, for example, plant material contains monosaccharides such as rhamnose and melibiose which are generally not found in mammalian tissues.

1.4.2 Complex sugars

Monosaccharides are covalently linked by a dehydration synthesis reaction, catalysed by a specific enzyme, to form either disaccharides such as sucrose (glucose–fructose) or lactose (galactose–glucose) (see *Figure 1.4*), or much longer, usually branching, sugar chains of variable length. *Oligosaccharides* consist of a few covalently linked monosaccharide units, and are often associated with proteins (glycoproteins or proteoglycans) or lipids (glycolipids); they are extremely heterogeneous in structure. *Polysaccharides* consist of many covalently linked monosaccharide units and have molecular weights of, sometimes, several millions of daltons; they are often comprised of monotonously repeating monosaccharide or disaccharide units, and usually have a structural or storage function. Glycogen and starch are examples of storage polysaccharides.

1.4.3 Glycoproteins and glycolipids

In cells, glycoproteins and glycolipids are usually membrane-associated or may sometimes be cytoplasmic. They may also be soluble and be present free in tissue fluid or serum, for example immunoglobulins are glycoproteins, as are some proteohormones. The sugar chains of glycoproteins and glycolipids are intimately involved in a whole range of cell–cell interactions, including phenomena as diverse as human and mammalian fertilization, plant pollination, embryogenesis, lymphocyte homing in infection and inflammation, plant–bacterial symbiosis and cancer metastasis. The structural diversity of oligosaccharides found on glycolipids and glycoproteins is enormous: some examples of oligosaccharides associated with glycoproteins are given in *Figure 1.5* to illustrate this point. There are subtle, or sometimes very obvious, alterations in the expression of sugar chains in several diseases such as rheumatoid arthritis and cancer. For these reasons, lectin histochemistry has been extensively used as an invaluable tool in mapping cellular glycosylation.

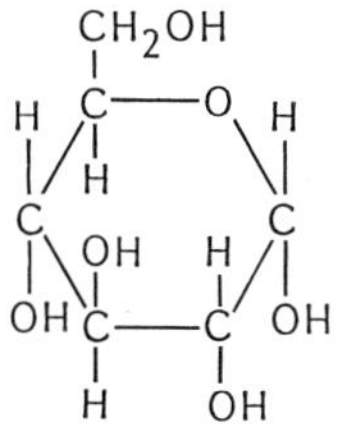

α-D-glucose $C_6H_{12}O_6$

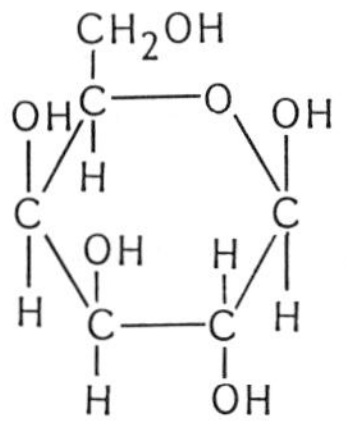

α-D-galactose $C_6H_{12}O_6$

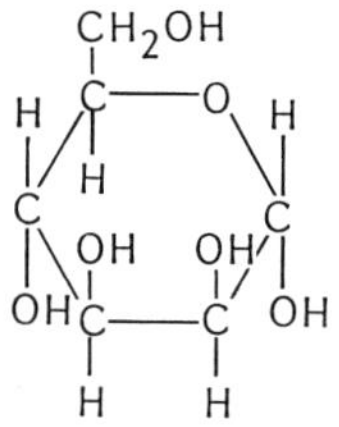

α-D-mannose $C_6H_{12}O_6$

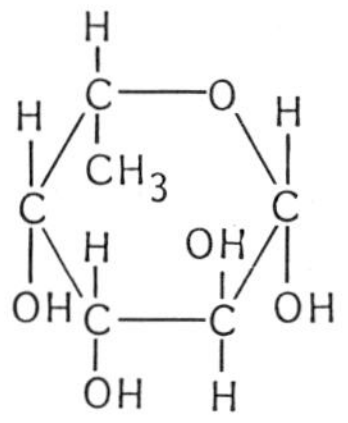

α-D-fucose $C_6H_{12}O_5$

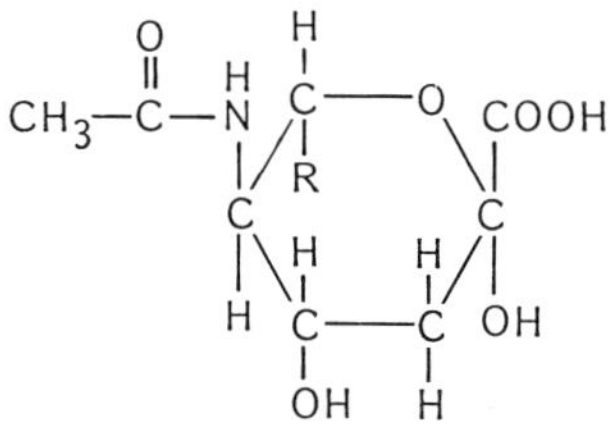

N-acetyl neuraminic acid $C_{11}H_{19}O_9$ (sialic acid)

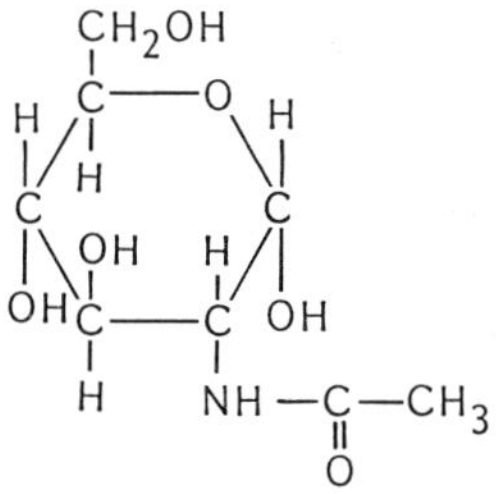

N acetyl glucosamine $C_8H_{15}O_6$

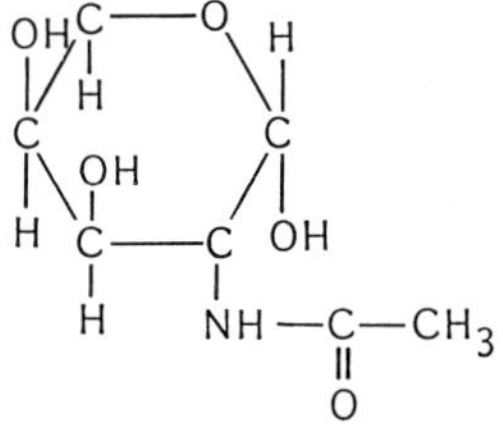

N acetyl galactosamine $C_8H_{15}O_6$

Figure 1.3: Structures of monosaccharides found in mammalian and human tissues.

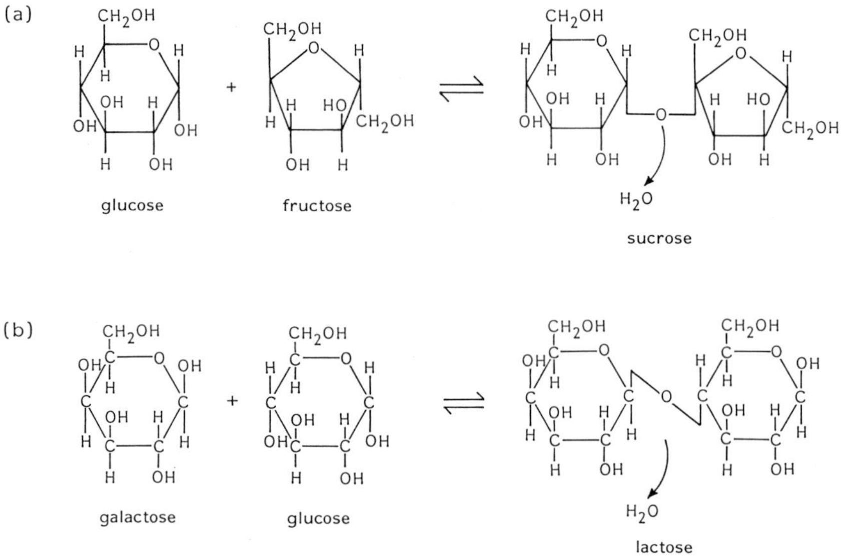

(a)

(b)

Figure 1.4: Dehydration synthesis reaction.
Two monosaccharides join by a glycosidic bond to form a disaccharide, with the expulsion of a molecule of water. (a) Glucose and fructose join to form a molecule of sucrose. (b) Galactose and glucose join to form a molecule of lactose.

1.4.4 Polysaccharides and glycosaminoglycans

Polysaccharides are linear or branching chains of repeating sugar units. They are often structural materials (e.g. cellulose in plants) or nutritional reservoirs (e.g. starch in plants, glycogen in animals. See *Figure 1.5*).

Glycosaminoglycans are unbranched polysaccharides of alternating uronic acid and acetylated aminosugar residues. Owing to their water-binding capacity, they are slimy and mucus-like in solution resulting in their viscosity and elasticity. They are usually found in extracellular matrix and have important structural properties.

Some examples of polysaccharides and glycosaminoglycans are given in *Table 1.2*.

1.5 Sugar specificity of lectins

The nominal specificity of a lectin is usually expressed in terms of the simple monosaccharide which best inhibits its effect. This has been most often demonstrated through inhibition of lectin-mediated cell

(a) Examples of O-linked oligosaccharides

```
gal — glcNAc
            \
             gal — glcNAc — gal — galNAc — ser (thr)
            /
gal — glcNAc
```

```
gal — glcNAc — gal — glcNAc
                            \
                             galNAc — ser (thr)
                            /
                         gal
```

```
gal — glcNAc
            \
             gal — glcNAc — galNAc — ser (thr)
            /
gal — glcNAc
```

```
gal — glcNAc
            \
             galNAc — ser (thr)
            /
gal — glcNAc
```

Examples of N-linked oligosaccharides

```
                                         fuc
                                          |
NANA — gal — galNAc — man                 |
                         \                |
                          galNAc — man — galNAc — galNAc —
                         /
NANA — gal — galNAc — man
```

```
NANA — gal — galNAc
                   \                       fuc
                    man                     |
                   /   \                     |
NANA — gal — galNAc      man — galNAc — galNAc —
                   /
                man
```

```
man — man
         \
          man
         /   \
man          man — galNAc — galNAc —
         \   /
          man
         /
man — man
```

(b) Structure of a polysaccharide, e.g. glycogen–repeating glucose
residues form a huge, branching complex polymer

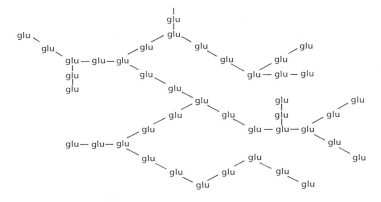

Figure 1.5: The structure of some complex sugars.
(a) The structure of glycoprotein and glycolipid oligosaccharide chains are highly diverse.
The oligosaccharides of glycoproteins may be N-linked or O-linked. Examples of both types
are shown. (b) Structure of a representative polysaccharide, glycogen. Repeating glucose
residues form a huge, branching complex polymer.

agglutination. For example, Lima bean (*P. lunatus*) lectin is said to be specific for galNAc because Lima bean lectin-induced haem-agglutination of blood group A erythrocytes is most effectively inhibited by the presence of the monosaccharide galNAc. However, it is extremely important to realize that saying that a lectin is specific for a particular monosaccharide is actually an over-simplification. Lectins *will* combine with monosaccharide moieties, and monosaccharides *will* inhibit lectin-induced agglutination; however, the combining site of the lectin is usually far more complex than this simple inhibition test would suggest. The actual structure recognized by the binding site of the lectin when it combines with its natural ligand is generally larger and more complex than a single monosaccharide. It is thought to involve typically three mono-

Table 1.2:   Examples of structural and storage polysaccharides and glycosaminoglycans

Polysaccharides	Occurrence	Composition
Structural polysaccharides		
Cellulose	Plants Outer mantles of tunicates	Linear polymer of up to 15 000 glc residues
Chitin	Exoskeleton of invertebrates Cell walls of fungi and algae	Linear polymer of glcNAc residues
Storage polysaccharides		
Starch	Energy reserve in plants	Branching polymer of glc residues
Glycogen	Energy reserve in animals and humans	Branching polymer of glc residues
Glycosaminoglycans		
Hyaluronic acid	Biological shock absorber/ lubricant in extracellular matrix in animal and human tissues Binding partner for cell adhesion molecules	Polymers of 250–250 000 glucuronic acid–glcNAc dimers
Chondroitin sulphate	Extracellular matrix of cartilage in animal and human tissues	50–1000 sulphated glucuronic acid–glcNAc residues
Dermatan sulphate	Extracellular matrix of skin in animal and human tissues	50–1000 sulphated glucuronic acid and iduronate–galNAc residues

saccharides, terminal and sub-terminal in the oligosaccharide chain, in a particular spatial arrangement; and sometimes even part of the protein or lipid to which the oligosaccharide is attached. In addition, hydrophobic and electrostatic interactions which are not located at the sugar binding site may also play a role in lectin binding to tissue structures. The principles involved in the binding of a lectin to its naturally occurring binding partner are illustrated in *Figure 1.6*. Two lectins with identical monosaccharide specificities, for example, two gal-specific lectins, may actually recognize very different sugar structures and give surprisingly different results in histochemistry. Binding constants between lectin–oligosaccharide are in the range 10^5–10^7 M^{-1} and between lectin–monosaccharide from 10^3–5×10^4 M^{-1}. Thus binding inhibition by a monosaccharide under experimental conditions will commonly require concentrations 100–1000 times higher than that of the preferred di- or trisaccharide, as illustrated in *Figure 1.7*.

The specific, natural oligosaccharide ligand is not yet known, as far as we know, for any lectin, although disaccharide (e.g. galβ1–3galNAc for PNA, see *Figure 1.8*) and trisaccharide (e.g. glcNAcβ1–4glcNAcβ1–4glcNAc for WGA and DSA, see *Figure 1.9*) structures that bind in preference to monosaccharide have been described for some lectins.

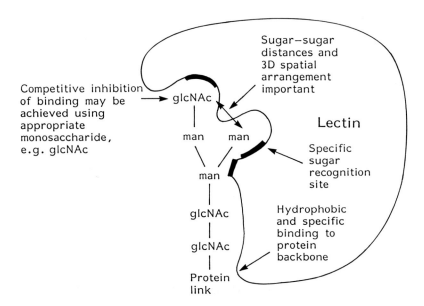

Figure 1.6:   Lectin binding to its natural binding partner.

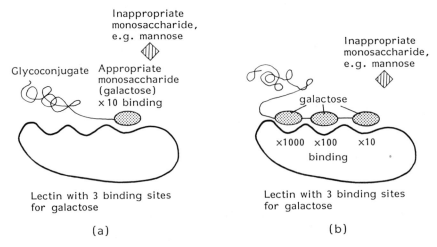

Figure 1.7: Lectin binding to monosaccharide and to its natural complex sugar binding partner.

(a) A lectin with a combining site specific for three gal sugars will bind to a single gal, and will do so about 10x more avidly than it would to an inappropriate monosaccharide (e.g. man). (b) The lectin will bind to its natural binding partner (three gals in the appropriate 3-D spatial arrangement) about 1000x more avidly. Binding will be inhibited by the presence of sufficient free gal monosaccharide in solution.

1.6 Sources and occurrence

Many of the better known and most intensively studied lectins are derived from plants. This is in part for historical reasons, since in their search for new lectins many early workers turned to plants

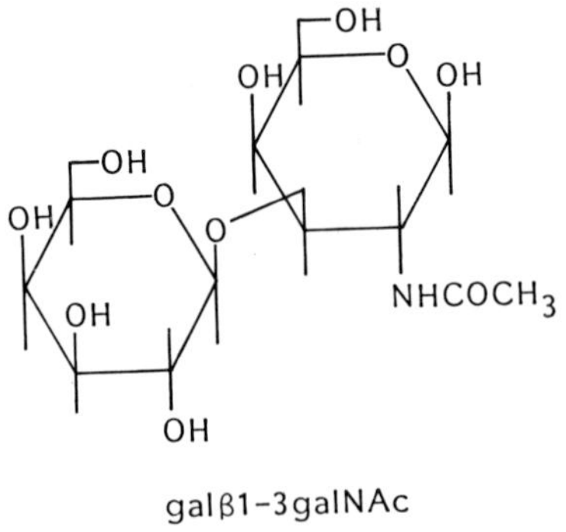

galβ1–3galNAc

Figure 1.8: Disaccharide binding partner for PNA.

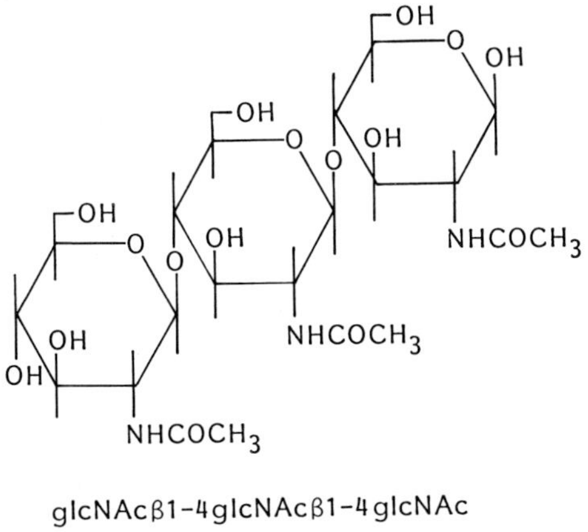

glcNAcβ1–4glcNAcβ1–4glcNAc

Figure 1.9:   Trisaccharide binding partner for WGA and DSA.

owing to their original discovery in plant material and ease of availability. Plant material, especially the seeds of the Leguminosae, Euphorbeaceae and Solanaceae families, are enormously rich sources of lectins (e.g. as much as 5% of the dry weight of peas, beans and lentils is lectin). Lectins have, however, subsequently been detected and isolated from diverse sources including viruses, bacteria, fungi, invertebrates, vertebrates, mammals and even in man.

It seems likely that virtually all living organisms contain some form of lectins. Sometimes, several lectins with different sugar binding specificities have been isolated from a single source; these are usually referred to as isolectins. A good example is *Bandeiraea simplicifolia* (also known as *Griffonia simplicifolia*), which has at least seven isolectins (called BSA-I, BSA-I-B4, BSA-I-AB3, BSA-I-A2B2, BSA-I-A3B, BSA-I-A4 and BSA-II), each with slightly different sugar binding characteristics. Similarly, *Ulex europeaus*, the gorse or furze, has at least two isolectins, UEA-I which binds α-L-fuc, and UEA-II which binds D-glcNAc. When we refer to 'peanut lectin' or 'soyabean lectin', we are probably referring to the isolectin which occurs at greatest concentration in that organism – other isolectins may also be produced but in lower concentrations. Some studies indicate that at different times of the growing season different isolectins may be produced in relatively greater or lesser proportions; this phenomenon has been put forward to explain variability in experimental results.

1.7 Detection

The most important property of a lectin, stated in its definition, is the ability to bind to carbohydrates and hence agglutinate cells. The agglutination is due to the fact that the lectin has at least two binding sites and it is therefore able to cross-link cells through interaction with carbohydrates on the cell membrane. Most investigators have detected lectins by agglutination experiments using human and animal erythrocytes and/or other cells.

All lectins have a more or less well defined binding specificity; however, some carbohydrate structures are more common in nature than others. Some lectins, which are specific for commonly occurring carbohydrate structures (e.g. glcNAc), therefore agglutinate cells of different blood groups or from several different species; other lectins, recognizing more unusual carbohydrate structures (e.g. fuc) appear to be more selective and may preferentially agglutinate cells from a particular animal species or agglutinate erythrocytes of a particular blood group only.

An alternative method of lectin detection is through their cross-linking and precipitation of naturally occurring or synthetic polysaccharides and glycoconjugates. This may be carried out either in liquid in a capillary tube or within an agarose gel.

Details of how to carry out agglutination or precipitation tests to detect the presence of a lectin are given in Section 2.1.3.

1.8 Lectin structure

Lectins are proteins or glycoproteins, and by definition have more than one binding site (most lectins have many binding sites, some as many as six). The amino acid sequence and detailed three-dimensional conformation of only a few lectins has been elucidated. Probably the most intensively studied lectin is Concanavalin A (Con A), a lectin derived from the jack bean *Canavalia ensiformis*, which has a nominal sugar binding specificity for man and glc. At neutral pH, Con A is a tetrameric molecule comprising four identical subunits (M_r = 26 500) folded together. The association of the subunits is pH dependent; at pH <5.6, it exists as a dimer; at pH >5.6 as a tetramer. It is a metalloprotein; each subunit contains one Ca^{2+} and one Mn^{2+} ion, the presence of which are essential for carbohydrate binding function. *Figure 1.10* illustrates, in simple form, the structure of Con A.

Lectins from related sources appear to share some amino acid sequence homology (e.g. lectins derived from Leguminosae species have a high degree of sequence homology), giving rise to the concept of 'lectin families', but are very different from lectins from other, unrelated, sources. Intriguingly, lectins derived from human and mammalian tissues appear to show some homology with plant lectins

Concanavalin A
tetramer

Showing carbohydrate
binding sites close to
Mn and Ca positions

h = hydrophobic binding

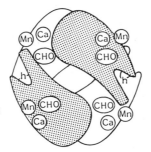

Figure 1.10:   The structure of Con A.

suggesting, perhaps, that they have been highly conserved through evolution.

1.9 Functions of lectins

The functions of lectins from diverse sources are gradually becoming understood. However, the role(s) of most lectins remains unclear. The high concentration of lectins in the seeds of legumes has given rise to speculation that they may have a controlling role in processes of germination or carbohydrate storage. Their role as plant defence proteins is now well established. Intriguingly many plant lectins have specificity for animal and not plant-derived complex carbohydrates and some lectins are highly toxic to animals. Intensive research is aimed at producing transgenic plants with lectin genes introduced to serve as a natural pesticide.

Early workers proposed that plant lectins may represent a primitive defence system and might be considered as 'plant antibodies'. Many studies have demonstrated that plant lectins can recognize and bind to complex carbohydrates on the surface of pathogenic microorganisms; however, the functional significance of this observation remains unclear. Perhaps of more relevance is the observation, made in several studies, that lectins can bind to and inhibit the growth of fungal hyphae, suggesting a protective role during germination and early seedling development.

From experiments such as the inhibition of sponge cell aggregation by carbohydrates and the development of sporing bodies by slime moulds, lectins appear to play a role in cell–cell recognition. Studies have also demonstrated the involvement of animal lectins in cell adhesion and recognition.

1.10 What can lectins be used for?

In simple terms, lectins can be used for all applications where monoclonal or polyclonal antibodies are commonly used. They have proved most useful in applications such as lectin histochemistry at light and electron microscope level to localize specific carbohydrate structures in cells and tissues (this topic is covered in Chapters 3, 4 and 5) and to visualize carbohydrate groups on glycoproteins separated by sodium dodecyl sulphate polyacrylamide gel electrophoresis (SDS–PAGE) or isoelectric focusing (IEF) and electroblotted on to nylon or nitrocellulose membranes (see Chapter 6). Interpretation and analysis of lectin binding is discussed in Chapter 7, and applications and usefulness of lectin binding studies is addressed in detail in Chapter 8. The final section of the book, Chapter 9, reviews the emerging field of endogenous (human or animal) lectins and their potential localization by histochemical techniques.

1.11 Why use lectins?

As stated in Section 1.10, lectins may be used for all applications where monoclonal or polyclonal antibodies are employed, so why use lectins, rather than antibodies? Lectins are a remarkable, huge, and amazingly diverse group of naturally occurring, specific, sugar binding molecules. For anyone interested in carbohydrate structures, lectins are the obvious tool for their investigation: few good antibodies directed against carbohydrates are available, and those that exist tend to be specific for small, well characterized sugars. Cells and tissues contain and express a vast, largely uncharted array of complex carbohydrates in the form of glycoproteins, glycolipids, glycosaminoglycans, etc. The essential roles of these carbohydrates in all manner of cell communication and cell signalling events is only just emerging in the exciting and rapidly expanding field of glycobiology. Lectins are the perfect tools for investigating these structures. There are currently well over 100 purified lectins commercially available (see Chapter 2), with diverse and usually only partially characterized sugar binding specificity, and the list is growing all the time. Their only major limitation is that, as yet, very little is known regarding their naturally occurring binding partners and the structures that they recognize on cells and tissues are heterogeneous and largely uncharacterized.

Lectins are wonderful tools for observing glycosylation changes associated with cell behaviour, development and disease. They have appeal for the histologist/cytologist (to distinguish cell populations), for the biologist (relating changes to cell behaviour), for the pathologist (detecting early changes in disease), and for the biochemist (in purifying glycoconjugates). Owing to their selective binding, plus simple competitive inhibition by monosaccharides, together with their relative cheapness, they are extremely useful for detection and isolation of glycoconjugates by affinity chromatography. After their detection by histochemistry, gel precipitation or Western blotting, single glycoproteins can be isolated from tissue extracts or tissue culture media by sequential lectin affinity chromatography columns. The oligosaccharides can then be cleaved off the glycoprotein by hydrazine, β-elimination or a glycanase for sugar sequencing, and the protein core itself can be sequenced.

References

Boyd WC, Raguera RM. (1949) Haemagglutinating substances for human cells in various plants. *J. Immunol.* **62**, 333–339.

Boyd WC, Shapleigh E. (1954) Separation of individuals of any blood group into secretors and non-secretors by use of a plant agglutinin (lectin). *Blood* **9**, 1195–1198.

Goldstein IJ, Hughes RC, Monsigny M, Osawa T, Sharon N. (1980) What should be called a lectin? *Nature* **285**, 66.

Kocourek J. (1986) Historical background. In: *The Lectins: Properties, Functions and Applications in Biology and Medicine* (eds IE Liener, N Sharon, IJ Goldstein). Academic Press, New York, 1–32.

Landsteiner K, Raubitschek H. (1907) Beobachtungen über Hämolyse und Hämagglutinin. Zentralb. Bacteriol. Parasitenkd. Infectionskr. *Hyg. Abt. 1: Orig.* **45**, 660–667.

Landsteiner K, Raubitschek H. (1909) *Biochem Z.* **15**, 33–51. Quoted by **Kocourek J.** (1986).

Morgan WTJ, Watkins WM. (1959) The inhibition of the haemagglutinins in plant seeds by human blood group substances and simple sugars. *Br. J. Exp. Pathol.* **34**, 94–103.

Peumans WJ, Van Damme EJM. (1994) Proposal for a novel system of nomenclature of plant lectins. In: *Lectins: Biology, Biochemistry, Clinical Biochemistry*, Vol. 10 (eds E Van Driessche, J Fischer, S Beeckmans, TC Bog-Hansen). Van Damme Textop, Hellerup. pp. 105–117.

Stillmark H. (1888) Thesis. University of Dorpat, Dorpat (Tartu). Quoted by Boyd WC. (1963) The lectins: their present status. *Vox Sang.* **8**, 1–32.

Watkins, Morgan (1952) Quoted by Kocourek J (1986) Historical background. In: *The Lectins: Properties, Functions and Applications in Biology and Medicine* (eds IE Liener, N Sharon, IJ Goldstein). Academic Press, New York.

2 Sources of Lectins and Other Reagents for Histochemistry

2.1 Isolation of lectins from plant material

2.1.1 Crude extraction

Plant material is by far the richest and most convenient source of lectins, although it is important to remember that the tissues and fluids of invertebrates, vertebrates and mammals as well as bacteria and fungi are also suitable sources.

The seeds of plants (in particular those of the Leguminosae, Solanaceae and Euphorbeaceae) are rich in lectins, and are easy and convenient sources for their extraction. In addition, the stems, leaves, bark and flowers of most plants can also contain appreciable amounts.

Lectin extraction can be as simple as soaking seeds in water or buffer for a few hours at room temperature. For example, soaking raw and unprocessed Lima beans obtained from a healthfood store in water for 1 h releases the lectin (*Box 2.1*), which can be readily detected by haemagglutination or immunoprecipitation, as described in Section 2.1.3. However, isolation usually takes the form of saline or buffer extraction of finely ground meal made from the seeds, or a fine mash or homogenate of other tissue. Pre-extraction with an organic solvent (e.g. methanol, acetone or diethyl ether) is sometimes useful at this stage to remove lipid. The lectin, and other proteins, will precipitate in the presence of the organic solvent and can be centrifuged out of suspension.

Stepwise ammonium sulphate precipitation (*Box 2.1*) may also be a useful preliminary clean-up step or a convenient way to concentrate

Figure 2.1: Ammonium sulphate precipitation.
Addition of ammonium sulphate to the sample causes precipitation of proteins (including lectins) which may then be separated from contaminants.

dilute plant extracts. Here ammonium sulphate is added to the crude extract to give progressively increasing levels of saturation (i.e. 10%, 20%, 30% (w/v) ammonium sulphate, etc.). At each stage the precipitated proteins are centrifuged out of suspension (see *Figure 2.1*) and the supernatant is tested for lectin activity. By this method a semi-purified lectin preparation (still, however, containing many contaminants) can be obtained.

The researcher may wish to experiment with a combination of different approaches, but a simple method for obtaining a semi-purified preparation of Lima bean (*P. lunatus*) lectin is given in *Box 2.1*. The method may be adapted for purification of lectin from other sources.

Box 2.1: Preparation of semi-purified Lima bean lectin

Lima beans may be purchased from most good healthfood stores.

1. Place 100 g Lima beans in a beaker. Add distilled water to just immerse the beans.
2. Leave the beans to soak for 30 min to 1 h (or overnight, if more convenient), with gentle occasional stirring.
3. Draw off the supernatant.
4. Stir ammonium sulphate into the liquid to give a 10% (w/v) solution. Leave the solution to settle for 10 min. Pour off the supernatant into a clean beaker, leaving a small amount of precipitated protein behind. Retain this.
   [It may be more efficient to centrifuge the solution at low speed (approximately 200–400 *g*) in a bench-top centrifuge for about 10 min to effectively separate the supernatant from the precipitate (see *Figure 2.1*).]
5. Stir ammonium sulphate into the liquid to give a 20% (w/v) solution. Leave the solution to settle for 10 min, or centrifuge as described in step 4. Pour off the supernatant into a clean beaker, leaving a small amount of precipitated protein behind. This is also retained.

(continued)

---

6. Repeat the process to precipitate proteins at 30%, 40%, 50%, 60%, 70% and 80% (w/v) ammonium sulphate.
7. Dissolve each precipitate in 5 ml distilled water.
8. Test each sample for lectin activity by haemagglutination of blood group A red blood cells (see *Box 2.5* for method).

2.1.2 Purification

Most modern lectin purification methods depend on affinity chromatography, which exploits the sugar binding capacity of the lectin. This may be sufficient alone, or may be performed in combination with one or more other conventional chromatographic techniques (e.g. gel filtration, ion exchange chromatography, etc.). A carbohydrate ligand that the lectin will recognize (either a monosaccharide, a polysaccharide or a complex natural or synthetic glycoconjugate) is immobilized on to a gel matrix (usually Sepharose or agarose beads). Many carbohydrates may be purchased commercially already immobilized on beads. If suitable beads are not available, it is quite straightforward to immobilize ligands on agarose beads in the laboratory: a suitable method is given in *Box 2.2*. The crude lectin-containing extract is passed down a column of the beads, and the lectin binds to the sugar and becomes adsorbed. Impurities can be washed away. Adsorbed lectin is then released by adding a solution of a simple or complex competing sugar, or by changing conditions of pH or ionic strength of the buffer bathing the column, or by adding a denaturing substance. This method is illustrated in *Figure 2.2*.

It is recommended that the worker consults the extensive literature on the subject for methods optimized for specific lectins of interest (e.g. see Porath and Kristiansen, 1975, for an excellent review), however, a simple method for purification of Lima bean (*P. lunatus*) lectin by affinity chromatography is given in *Box 2.3*.

Box 2.2: Immobilization of (neo)glycoprotein on agarose beads

All solutions should be pre-cooled to 4°C before use.

Coupling buffer. 0.1 M NaHCO$_3$, pH 8.3, with 0.5 M NaCl.

Swelling and washing the gel.
1. Allow 2 g cyanogen bromide-activated sepharose 4B beads (see note 1) to swell for 15 min in 20 ml 1 mM HCl, with continuous agitation in an end-over-end mixer.
2. Wash the beads 10 times in 20 ml aliquots of 1 mM HCl; after each wash, allow the beads to settle, and pour off the supernatant and finings (note 2).

(*continued*)

Preparation of the ligand to be coupled.
1. For each 1 ml gel, prepare 2 ml solution of the carbohydrate ligand (note 3) at a concentration of 5–10 mg ml⁻¹ in coupling buffer (notes 4 and 5).

Coupling of ligand to beads.
1. Briefly wash the gel in coupling buffer, then transfer it immediately to the solution of carbohydrate ligand. The ratio of volumes should be: 1 part gel:2 parts carbohydrate ligand solution.
2. Gently mix the gel and carbohydrate ligand together in an end-over-end mixer for 2 h or overnight.

Blocking excess reactive groups.
1. Allow the beads to settle and discard the supernatant.
2. Wash the beads extensively, with gentle agitation in an end-over-end mixer, with Tris-buffered saline (TBS), pH 7.6 (see Appendix A for recipe).

Washing the product.
1. Wash the beads by alternate cycles of: (a) coupling buffer, and (b) sodium acetate buffer, pH 4.5 (see Appendix A for recipe), both incorporating 0.5 M NaCl (note 6).

Notes
1. This method may be used to couple any protein ligand to beads for affinity chromatography. Cyanogen bromide reacts with vicinol diols of agarose, dextran or cellulose to produce a reactive matrix that can be subsequently derivatized with either spacer molecules or ligands containing primary amines. Cyanogen bromide is an extremely hazardous substance and should be handled with great care. However, cyanogen bromide-activated beads may be purchased ready to use from suppliers such as Sigma and are perfectly safe to use on the laboratory bench without any special precautions.
2. The addition of HCl preserves the activity of the reactive groups which hydrolyse readily at high pH.
3. This method, as described in note 1, will couple ligands containing primary amines to the beads, therefore it does not work for simple monosaccharides or oligosaccharides. Monosaccharides, and more complex sugars, may be purchased from commercial suppliers linked to protein carrier molecules [most usually bovine serum albumin (BSA); these compounds are sometimes referred to as neoglycoproteins]. They are relatively inexpensive to buy, and are ideal for this method. An alternative approach is to use glycoproteins carrying the appropriate monosaccharide or oligosaccharide of interest.
4. Coupling is most efficient at pH 8–10, where amino groups of the ligand to be coupled are in an unprotonated form, and in the presence of high concentrations of salt, which minimizes protein–protein adsorption caused by the polyelectrolyte nature of proteins.
5. Tris and buffers containing amino groups must *not* be used, as they will couple to the gel.
6. This step ensures that no free carbohydrate ligand remains ionically bound to immobilized carbohydrate or gel. Protein desorption occurs only when the pH is changed. pH changes do not cause loss of covalently bound carbohydrate ligand.
7. Some beads [e.g. agarose (gal polymers), Sepharose (gal), Sephadex (glc)] are suitable as matrices for affinity isolation, either as supplied or after mild hydrolysis.

Figure 2.2:   Principle of affinity chromatography to purify lectin from a crude extract.
(a) Crude extract applied to a column of appropriate sugar immobilized on beads. (b) Lectin binds to sugar on beads; impurities are eluted to waste. (c) A solution of appropriate sugar is applied to the column. (d) Lectin is displaced from the beads and collected.

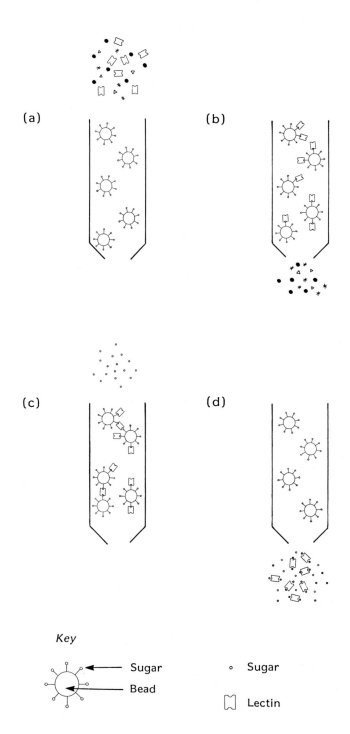

(a)

(b)

(c)

(d)

Key

Sugar

Bead

Sugar

Lectin

Impurities in crude extract

Box 2.3: Purification of Lima bean lectin by affinity chromatography

1. Pool the fractions which gave maximum haemagglutination in the method described in *Box 2.1*.
2. Pour galNAc immobilized on agarose beads (Sigma) into a small chromatography column.
3. Wash the column and equilibrate it with 5× column volumes of lectin buffer (see Appendix A for recipe).
4. Apply semi-pure Lima bean lectin extract to the column and allow it to interact with the column for about 30 min at room temperature (note 1). Wash unbound material with 5× column volumes of lectin buffer. Collect unbound material.
5. Apply 2× column volumes of freshly prepared 0.1 M galNAc solution in lectin buffer to the column (note 2). Collect the eluted material.
6. Wash the column and re-equilibrate it with lectin buffer.
7. Dialyse the material collected in step 6 (i.e. lectin) overnight against lectin buffer to remove excess monosaccharide (see *Box 2.4*).

Notes
1. In our experience, the yield of purified lectin is markedly improved by allowing the extract to react with the column for at least 30 min. This can be most simply done by capping the bottom of the column and leaving it to stand at room temperature before proceeding with the method. If a more sophisticated chromatography set-up is used, it may be possible to recirculate the extract through the column using a pump. We have sometimes allowed an extract to recirculate overnight at 4°C with significant improvement in yield.
2. Most solutions of carbohydrates are stable at room temperature or 4°C; galNAc and glcNAc are exceptions and solutions should be prepared fresh daily.
3. Unbound material (step 4) and lectin (step 5) after dialysis (see *Box 2.4*) may be tested for lectin activity by haemagglutination of blood group A red blood cells as described in *Box 2.5*.

Troubleshooting
1. Unbound material (step 4) contains lectin activity. This is to be expected. Not all the lectin in solution will be adsorbed on to the column, but most will. Adsorption can be increased by circulating the solution through the column several times or allowing it to react with the column for a longer period of time (see note 1). A slow flow rate through the column will also increase lectin adsorption. If very little lectin is being retained by the column, this may indicate (particularly if the column has been used many times) that the column is dirty (i.e. that either the carbohydrate residues on the column are already bound with lectin from previous runs, or bound non-specifically to contaminants from previous runs). The column should be regenerated by following the manufacturer's instructions or replaced by new gel.
2. Lectin is adsorbed on to the column, but is not being released by the sugar solution. This can sometimes be a problem with affinity chromatography; the lectin often binds with greater avidity to immobilized monosaccharide (i.e. on the beads) than to free monosaccharide in solution. Increasing the concentration of the galNAc solution to as much as 0.5 M may help, as may increasing the NaCl concentration in the lectin buffer to up to 0.5 M. A more aggressive method is to wash the column with alternating cycles of 0.5 M NaCl in lectin buffer at pH 2 followed by pH 12. This should be done quickly and the column re-equilibriated with lectin buffer pH 7.6 as soon as possible to avoid damaging it.

Box 2.4: Dialysis

Dialysis membrane can be purchased from a number of commercial suppliers (e.g. Sigma) in several different forms. The most convenient form is probably dialysis tubing, purchased in a roll, of diameter approximately 2 cm. Dialysis tubing is available with various molecular weight cut-offs (i.e. pore size). Obviously, tubing should be used with an appropriate pore size for the job. For dialysing lectin to remove contaminants such as salts and monosaccharides, dialysis tubing with a molecular weight cut-off of approximately 12 000 Da is ideal. The experimental set up for dialysis of a sample against buffer or water is illustrated in *Figure 2.3*.

1. Dialysis tubing should be prepared exactly according to the manufacturer's instructions. This can be done in advance, and convenient lengths of tubing stored in buffer in the fridge ready for use.
2. Cut a suitable length of dialysis tubing – it must be long enough to hold the volume of the sample to be dialysed, with capacity for in-flow of water or buffer during dialysis, plus enough extra length to seal both ends. As a rough guide, one should allow at least twice the length of tubing required to hold the volume of solution to be dialysed.
3. Carefully seal one end of the tubing. Special dialysis clips are available for this purpose, but tying a strong double or triple knot in the end of the tubing is perfectly adequate.
4. Introduce the sample into the tubing; expel air bubbles.
5. Seal the other end of the tubing, as above, allowing plenty of empty space for buffer or water to flow into the tube during dialysis.
6. Place the whole 'sausage' in a copious amount of water or buffer in a large beaker or conical flask. As a rough guide, 50 ml of solution should be dialysed against at least 500 ml of water or buffer.
7. Place a magnetic stirring bar in the bottom of the beaker or flask and stand it on a magnetic stirrer. Set the stirrer at a low speed so that the buffer or water is stirred very gently and continuously while dialysis takes place.
8. Allow dialysis to proceed for at least 12–24 h, preferably at 4°C, and during that time change the water or buffer at least two or three times.

2.1.3 Testing

Haemagglutination. Lectins were first detected by their ability to agglutinate erythrocytes, and probably the easiest and most convenient method of detecting lectin activity remains agglutination of a panel of human and/or animal red blood cells, or other cells. Red blood cells digested with papain, trypsin, neuraminidase or other enzymes may also be employed. Only a small number of lectins recognize the common human blood group sugars, and enzyme treatment reveals a range of sub-terminal sugar residues, making the cells susceptible to agglutination by a wider range of lectins. Red blood cells suitable for haemagglutination experiments may be purchased from a number of commercial suppliers (e.g. Sigma).

Figure 2.3: Dialysis tubing containing the sample is dialysed against a large volume of buffer or water.

Lectins may be of different types:

- lectins that agglutinate cells without regard to their origin, species or blood group,
- lectins that preferentially agglutinate one cell type, or erythrocytes from one particular species,
- blood group-specific lectins.

It is helpful when assaying for the presence of a lectin by agglutination to know what sort of cells might be appropriate (e.g. if the lectin has a monosaccharide specificity for galNAc, it should clump human blood group A erythrocytes). If this information is not available, a wide range of different cells should be tested, or an alternative method sought.

If seeking a suitable lectin for a specific purpose, for example to identify a particular cell type or function, it is obviously a good idea to screen for agglutination of that cell type rather than to use erythrocytes. For example, if one is seeking a lectin to label or identify lung cancer cells, it would be appropriate to seek agglutination of a suspension of lung cancer cells rather than any other cell type.

The technique of haemagglutination is very quick, simple and straightforward. It is described in *Box 2.5*; suitable apparatus for carrying out haemagglutination is illustrated in *Figure 2.4*, and the results of haemagglutination in *Figure 2.5*.

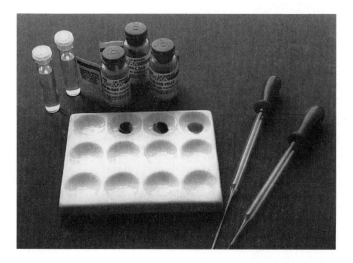

Figure 2.4:   Simple equipment for haemagglutination.

Box 2.5: Testing for lectin activity by haemagglutination

1. Place a drop of red blood cells (or other cells, as appropriate) on to a porcelain tile or microscope slide using a Pasteur pipette (see *Figure 2.4*) (note 1). The cells should be in a 4% (v/v) suspension in a medium free from potentially competing sugar. Lectin buffer, which contains the calcium and magnesium ions that many lectins require (see Appendix A for recipe), is ideal (note 2).
2. Add a drop of lectin solution to the cells (again, a solution prepared in lectin buffer is ideal) and tilt the tile or slide to mix for a few seconds.
3. Set up a negative control consisting of a drop of cells mixed with a drop of lectin buffer for comparison with the test, and a positive control of cells mixed with a known lectin preparation (e.g. a commercial preparation at a concentration of approximately 100 µg ml^{-1} in lectin buffer, or a previous, successful in-house preparation).
4. Agglutination may occur virtually instantaneously. Strong agglutination will be readily visible with the naked eye (cells will obviously and visibly clump, as in clotting). Moderate agglutination can be readily observed by viewing the cells under a light microscope at low power ($\times 4$ or $\times 10$) (cells will be seen to aggregate in very large clumps with only scanty single cells in the background). This is illustrated in Figure 2.5. Sometimes agglutination will take longer, and cells should be allowed to incubate with the lectin for at least 30 min before it can be concluded that haemagglutination has not taken place (note 3).

Notes
1. If a very large number of lectins/extracts are to be tested a 96-well microhaemagglutination plate may be preferable.

(*continued*)

2. A range of human erythrocytes are commercially available ready to use for haemagglutination from a number of suppliers including Sigma. In our experience, this is an economical and convenient way of getting hold of suitable cells. They are stable at 4°C for several weeks.

3. If cells are allowed to incubate with lectin for any length of time it is essential that the mixture is not allowed to dry out. Incubation in a moist chamber is essential. We would recommend placing the microscope slide or tile in a sealed humid container, for example a Petri dish or sandwich box lined with wetted filter paper.

Troubleshooting

This technique is usually straightforward; however, if agglutination fails it is probably due to one of the following reasons.

1. Incorrect cell:lectin ratio. Agglutination will be successful within wide limits of concentration ratios, however a very low lectin titre or very low cell number may be insufficient for successful agglutination to take place. Too high a lectin concentration may also inhibit agglutination. Try adjusting the concentration of lectin and cells.

2. Old cells. Cells that have been stored for any length of time may deteriorate. Ideally, freshly drawn cells should be used. In our experience, however, commercially available red blood cells supplied in the presence of preservatives will remain stable for weeks if stored in the refrigerator.

3. Competing sugars. The presence of competing sugars will inhibit agglutination. This applies not only to monosaccharides, but especially to glycoconjugates such as glycoproteins and glycolipids. For example, many serum components are heavily glycosylated: haemagglutination should always be carried out under serum-free conditions. If in doubt, cells should be washed gently in lectin buffer prior to testing.

Precipitation. A more refined screening procedure depends on the ability of lectins to precipitate glycoconjugates. The most straightforward way of doing this is in the medium of an agarose gel. This test is called Ouchterlony gel diffusion; suitable apparatus for Ouchterlony gel diffusion is illustrated in *Figure 2.6,* an example of results is shown in *Figure 2.7* and a method is given in *Box 2.6.* Formation of a precipitate indicates multivalent lectin activity. Moreover, it can provide information regarding lectin specificity and anomeric preference.

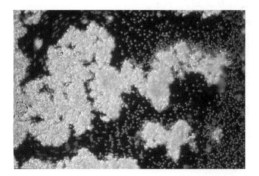

Figure 2.5: Group A red blood cells agglutinated by the addition of DBA.

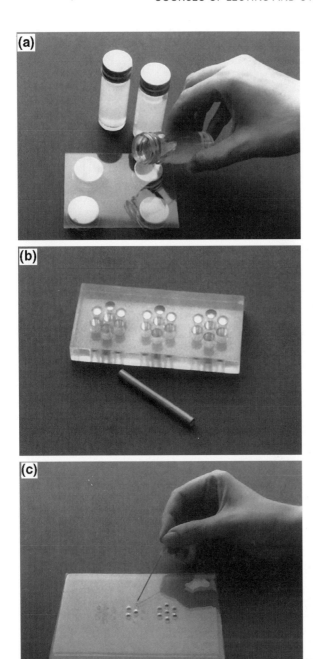

Figure 2.6:   Ouchterlony gel diffusion.
(a) Pouring the gel. (b) Hole punch. (c) Preparation of wells.

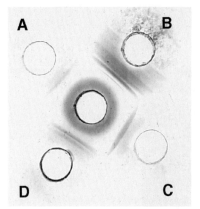

Figure 2.7: Ouchterlony gel diffusion.
The centre well contained human serum, wells A–D contained crude plant extracts rich in lectin. Where lectin has combined with serum glycoconjugates, discrete bands of precipitate (stained with CBB) are seen.

Box 2.6: Ouchterlony gel diffusion (see *Figures 2.6* and *2.7*)

1. Prepare a 1.5–3% (w/v) agarose gel solution by heating agarose powder in lectin buffer, with continual stirring, until the agarose melts (note 1).
   NB. Take care as the agarose will char easily. Use of a boiling waterbath or domestic microwave oven is preferable to direct heat.
2. Allow the agarose solution to cool to about 40°C, with occasional stirring (note 2).
3. Meanwhile, balance a clean, dry sheet of glass on supports (e.g. a bottle cap under each corner, see *Figure 2.6*). Check that the glass sheet is absolutely level using a spirit level and make any corrections by tucking folded paper under the supports, as appropriate.
4. Pour the gel over the plate in a rapid, smooth, even motion to give a uniform layer approximately 5 mm thick. This should be done with care as the gel will begin to set almost as soon as it hits the cold glass – the technique improves with practice! If you have problems, it may help to warm the glass plate as this will delay the setting of the gel.
5. Allow the gel to set at room temperature for about 30 min.
6. Punch two wells through the entire thickness of the gel, about 5–10 mm apart. Special punches are available for this purpose (*Figure 2.6*), but a home-made one will do (e.g. a short length of metal tubing approx. 5 mm in diameter, or the plastic casing of a biro, or the end of a pasteur pipette). Winkle out the plug of gel using a pin or syringe needle.
7. Fill one well with lectin solution. Fill the second well with a solution of glycoconjugate; naturally occurring glycoproteins (e.g. fetuin), synthetic glycoconjugates, or heterogeneous mixtures (e.g. serum) are all suitable.
8. Cover the gel (e.g. place it in a sealed, humid sandwich box) and leave it at room temperature or 37°C, for 2–12 hours. Lectin and glycoconjugates will each diffuse through the gel matrix. When they meet, if they react, an insoluble precipitate will form (see *Figure 2.7*). If the precipitate is a substantial one, it will be visible with the naked eye as a fine white line in the colourless gel (best observed if the gel is tilted to the light against a dark coloured background). If a lighter precipitate is formed, it may be harder to see and require staining with Coomassie brilliant blue (CBB).

(continued)

9. Before staining the precipitate, any unprecipitated proteins must first be washed away. First dry the gel by placing several thicknesses of filter paper or paper towel on top of it, and weighting them down with any suitable heavy weight. Replace the filter papers hourly until the gel is dry. Next soak the gel in distilled water until it swells to its original thickness (approx. 30 min). Cover in filter papers and weight as before. Repeat the washing/drying cycle twice more.

10. Immerse the gel in CBB total protein stain for 30 min, then destain over a period of 24–48 h in several changes of destaining solution. Recipes are given in Appendix A. Precipitate should show up deep blue against a transparent backround.

Notes
1. We find it convenient to prepare the agarose in 30 ml glass Universal tubes. A tube is usually sufficient for setting up one Ouchterlony gel diffusion plate. Spare tubes can be stored for months at 4°C until required.
2. As a rough guide, the outside of the glass container should feel hot to the hand, but not uncomfortably so.
3. It may be convenient to set up multiple tests on one agarose gel. Rosettes of wells (e.g. a central well containing lectin and outer wells containing glycoconjugate) are very convenient. Templates for this purpose are available commercially (*Figure 2.6*).
4. It is a good idea to test a range of dilutions of lectin against a range of dilutions of glycoconjugate solution as the ratio of one to the other may be critical. Too high a concentration of lectin or glycoconjugate may prevent formation of a precipitate (a phenomenon called prozoning) as effectively as too low a concentration.

Troubleshooting
1. The most common reason for this form of precipitation failing is inappropriate ratios of lectin : glycoconjugate. The optimum ratio can only be determined by trial and error, and it is therefore a good idea to set up a wide range of dilutions of both lectin and glycoconjugate solutions.
2. It is essential to wash out unprecipitated protein prior to staining or 'dirty' background protein staining will obscure faint lines of specific precipitate.

2.2 Commercial sources of lectins and other reagents

2.2.1 Suppliers

Unless the investigator is interested in a new or very unusual lectin, it is probably easier and more economical in terms of both time and money to purchase crude or purified lectin preparations from a commercial source. An increasingly wide range of lectins is available commercially, and many are supplied ready labelled with enzymes, fluorochromes or biotin. A list of some commercially available lectins, with information about their binding characteristics, is given in *Table 2.1*.

A number of companies in the UK supply lectins (e.g. Sigma-Aldrich Chemical Co., EY Laboratories, Vector Laboratories and

continues on p. 37 →

Table 2.1: Range of commercially available lectins and their characteristics[a]

Source of lectin		Common abbreviation	Inhibitory carbohydrate	Blood group or other binding specificity
Latin name	Common name			
Abrus precatorius				
agglutinin (non-toxic)	Jequirity bean	APA	D-gal	
abrin A² (abrin b+c³)(toxin)			D-gal	
abrin C² (abrin a³)(toxin)			D-gal	
Aegopodium podagraria	Ground elder	APP	galNAc > lac > gal	
Agaricus bisporus	Mushroom	ABA	β-D-gal(1→3)D-galNAc	
Allomyrina dichotoma	Japanese beetle	Allo A	β-D-gal	
Amaranthus caudatus	Amaranthin	ACA		
Anguilla anguilla	Freshwater eel	AAA	α-L-fuc	H blood group
Arachis hypogaea	Peanut	PNA	β-D-gal(1→3)-D-galNAc	T antigen
Artocarpus integrifolia	Jack fruit	Jacalin, AIA	Methyl-α-D-galactopyranoside	T antigen, IgA1
Bandeiraea simplicifolia (also called *Griffonia simplicifolia*)		BSA	α-D-gal	
BS-I			α-D-gal	
BS-I-B$_4$			α-D-gal	
BS-I-AB$_3$			α-D-gal, α-D-galNAc	
BS-I-A$_2$B$_2$			α-D-gal, α-D-galNAc	
BS-I-A$_3$B			α-D-gal, α-D-galNAc	
BS-I-A$_4$			α-D-galNAc	
BS-II			D-glcNAc	
Bauhinia purpurea	Camel's foot tree	BPA	β-D-gal(1→3)-D-galNAc	
Bryonia dioica	White bryony	BDA	galNAc > lac > melibiose	
Canavalia ensiformis	Jack bean	Concanavalin A, Con A	α- methyl mannopyranoside < α-D-man > α-D-glc > glcNAc	

Species	Common name	Abbreviation	Specificity	Notes
Cancer antennarius	California crab	CCA	9-O-acetyl sialic acid, 4-O-acetyl sialic acid	
Caragana arborescens	Siberian pea tree	CAA	D-galNAc	
Cicer arietinum	Chick pea, ceci bean	CPA	Unknown	Fetuin and IgM
Codium fragile	Green marine algae	–	D-galNAc	
Colchicum autumnale	Meadow saffron	CA	lac > galNAc > gal	
Cytisus scoparius	Scotch broom	CSA	D-galNAc >> lac, melibiose > gal	
Datura stramonium	Jimson weed or thorn apple	DSA	Chitotriose > chitobiose >> glcNAc	
Dolichos biflorus	Horse gram	DBA	α-D-galNAc	Anti-blood group A1, and Cad
Erythrina corallodendron	Coral tree	ECorA	β-D-gal-(1→4)-D-glcNAc	
Erythrina cristagalli	Coral tree	ECA	β-D-gal-(1→4)-D-glcNAc galNAc > lac > glcNAc > gal	
Euonymus europaeus	Spindle tree	EEA	α-D-gal-(1→3)-D-gal	
Galanthus nivalis	Snowdrop	GNA	Non-reducing end of terminal α-D-mannosyl residues	Agglutinates rabbit erythrocytes
Glycine max	Soyabean	SBA	D-galNAc, gal	
Griffonia simplicifolia – see *Bandeiraea simplicifolia*				
Helix aspersa	Garden snail	HAA	D-galNAc >> glcNAc	Blood group A
Helix pomatia	Roman or edible snail	HPA	D-galNAc > glcNAc >> gal	Blood group A
Homarus americanus	Californian lobster	HMA	N-acetyl neuraminic acid, galNAc	
Iberis amara	–	IAA	Unknown. Not inhibited by simple sugars	Blood group M
Laburnum alpinum	Scotch laburnum	LAA	glcNAc	
Laburnum anagyroides	Gold chain	LAL		
Lathyrus odoratus	Sweet pea	LOA	α-D-man	
Lens culinaris (or *Lens esculenta*)	Lentil	LcH, LCA	α-D-man, D-glc	
Limax flavus	Cellar slug	LFA	N-acetyl neuraminic acid	

(continued)

Source of lectin		Common abbreviation	Inhibitory carbohydrate	Blood group or other binding specificity
Latin name	Common name			
Limulus polyphemus	Horseshoe crab	LPA, Limulin III	NeuAc	Glucuronic acid, phosphorycholine analogues
Lotus tetragonolobus (see *Tetragonolobus purpureas*)				
Lycopersicon esculentum	Tomato	LEA	D-glcNAc oligomers	
Maackia amurensis MAH MAL	Maackia	MAA	Unknown α-sialyl-(2-3)-gal	
Maclura pomifera	Osage orange	MPA	α-D-gal, α-D-galNAc melibiose	T lymphocytes
Mangifera indica	Mango	MIA	Unknown	
Momordica charantia	Bitter pear melon	MCA	D-gal, D-galNAc	
Mycoplasma gallisepticum	–	MGA	NeuAc	Glycophorin
Naja mossambica mossambica	Mossambica cobra venom	SVAM	Unknown	Heparin
Naja naja kaouthia	Kaouthia cobra venom	SVAK	Unknown	Heparin
Narcissus pseudonarcissus	Daffodil	NPA	Man	
Oryza sativa	Rice	OSA	Unknown	
Persea americana	Avocado	PAA	Unknown	
Phaseolus coccineus	Scarlet runner bean	PCA	Unknown	Fetuin
Phaseolus limensis (or *Phaseolus lunatus*)	Lima bean	LBA	D-galNAc, α-D-galNAc-(1→3) (α-L-Fuc-[1→2])-D-gal	Blood group A
Phaseolus vulgaris PHA-E	Red kidney bean	PHA	Complex carbohydrate	

		PHA-L	Complex carbohydrate	
		PHA-P	Complex carbohydrate	
		PHA-M	Complex carbohydrate	
Phytolacca americana	Pokeweed	Pokeweed mitogen, PWM or PWA	D-glcNAc oligomers	
Pisum sativum	Garden pea	PSA	α-D-man, α-D-glu	
Pseudomonas aeruginosa	–	PAA	D-gal	
PA-I			fuc, man	
PA-II				
Psophocarpus tetragonolobus	Winged bean	PTA	D-galNAc, D-gal	
Ptilota plumosa	Red marine algae	PPA	α-D-gal	Blood group B
Ricinus communis	Castor oil bean	RCA		
Toxin RCA60 (RCAII , ricin D or RCL III)			D-galNAc, β-D-gal	
Toxin RCA120 (RCAI or RCL I+II)			β-D-gal	
Robinia pseudoacacia	False acacia or black locust	RPA	Unknown	
Salvia horminum	Salvia	SHA	galNAc, complex carbohydrates	Tn, Cad
Salvia sclarea	Salvia	SSA	galNAc, complex carbohydrates	Tn, Cad
Sambucus nigra	Elderberry	SNA	Lactose (β-D-gal-(1→4)-D-glc) > gal > N-acetyl neuraminic acid	
Sarothamnus scoparius	–	SRA	D-glcNAc oligomers	
Solanum tuberosum	Potato	STA		
Sophora japonica	Japanese pagoda tree	SJA	β-D-galNAc, D-gal	Blood groups A and B
Tetragonolobus purpureas (or *Lotus tetragonolobus*)	Winged pea or asparagus pea	LTA	α-L-fuc	Blood group H
Trichosanthes kirilowii	China gourd, tianhuafen	TKA	lac > β-D-gal	
Trifolium repens	White clover	RTA	2-Deoxyglucose	

(continued)

Source of lectin				
Latin name	**Common name**	**Common abbreviation**	**Inhibitory carbohydrate**	**Blood group or other binding specificity**
Triticum vulgaris	Wheatgerm	WGA	β-D-glcNAc, β-D-glcNAc NeuAc, galNAc > lac > gal galNAC > lac > gal	
Tulipa sp.	Tulip	TL		
Ulex europaeus	Gorse or furze	UEA		
UEA-I			α-L-fuc	Blood group O/H
UEA-II			N,N'diecetylchitobiose	
Urtica dioica	Stinging nettle	UDA	D-glcNAc glcNAc-β-(1,4) glcNAc Oligomers of β(1,4) glcNAc	
Vicia faba	Fava bean or broad bean	VFA	D-man > D-glc > glcNAc	
Vicia graminea	–	VGA		Blood group N
Vicia sativa	–	VSA	D-glc, D-man	
Vicia villosa	Hairy vetch	VVA		
A4			D-galNAc	Blood group A1
A2B2			D-galNAc	Blood group A1 / Tn
B4			D-galNAc	Tn erythrocytes
Vigna radiata	Mung bean	MBA or VRA	α-D-gal	
Viscum album	European mistletoe	VAA		
ML-I			gal	
ML-II			gal, galNAc	
ML-III			galNAc	
Wisteria floribunda	Japanese wisteria	WFA	D-galNAc >> lac > gal	

[a] An increasingly wide range of lectins is available commercially. Many of these can be purchased ready labelled with enzymes, fluorescent tags or biotin. The list in this table is by no means comprehensive and, obviously, new products come on to the market all the time, but it gives an idea of the selection available. An excellent reference work is Pusztai *et al.* (in press).

Boehringer Mannheim). Details of the companies are given in Appendix II.

Sigma stocks a range of more than 50 lectins, including some unusual ones. They are available as both purified and crude preparations. Many lectins are available conjugated to peroxidase, biotin, fluorescein isothiocyanate (FITC), tetramethylrhodamine isothiocyanate (TRITC), colloidal gold, alkaline phosphatase, etc., and some are available immobilized on agarose or Sepharose beads for affinity chromatography. Sigma also stock polyclonal antisera raised in rabbits against a very limited range of lectins. Conjugation reagents to label lectins with FITC, peroxidase, biotin, etc. are also available.

EY Laboratories (UK agents: Bradsure Biologicals) stock lectins from around 70 sources, including a number of fairly obscure ones (delivery of the more unusual lectins can sometimes take several weeks). Polyclonal antisera raised in rabbits against some of the more common lectins are available. Most lectins are also supplied labelled with FITC, TRITC, Texas Red, biotin, peroxidase, alkaline phosphatase, ferritin, colloidal gold and immobilized on gels for affinity chromatography.

Vector Laboratories Ltd supply a range of approximately 30 lectins. Most are available labelled with biotin or fluorescent tags (FITC, TRITC or both) and many are supplied immobilized on agarose beads for affinity chromatography.

Boehringer Mannheim also stock a limited range of less than 20 lectins, both in native form and labelled with digoxygenin.

2.2.2 Cost of commercially available lectins

Prices vary, but on average fall in the range of £10–50 per mg of native, unlabelled lectin. Labelled lectins are more expensive than native ones. Polyclonal antisera against them costs on average about £30 per ml. For most uses (e.g. histochemistry or lectin blotting using direct or indirect visualization techniques), a working dilution of lectin of approximately 1–10 µg ml^{-1} is usual, so a little goes a long way.

2.3 How to make your own polyclonal antibodies against lectins

Antisera against all but the most commonly used lectins are generally unavailable commercially. If a large amount of antibody against a specific lectin is required, it is relatively simple (if one has access to appropriate facilities) to produce good quality polyclonal antisera to lectins, for moderate cost, 'in house'. Of course, in order to do this Project and Personal Licenses must be applied for through the Home Office.

Owing to the time, trouble, and specialist facilities required to raise antibodies, unless the investigator has easy access to specialist facilities and experience of animal work, labelling lectins with biotin (described in *Box 2.8*) is probably a more sensible option. Biotinylation is quick, cheap and easy, and biotin-labelled lectins give excellent results in histochemistry.

Detailed description of theory and methodology for antibody production lies beyond the scope of this book – the reader is referred to any good manual (Harlow and Lane, 1988, is an excellent reference guide) – but theoretical considerations specific to producing antibodies against lectins, and examples of tried and tested methods are given below.

2.3.1 General protocol

For practical reasons, rabbits represent a good choice for the production of polyclonal antisera. Around 500 ml of immune serum can be harvested from one animal during the course of an immunization regime without harming the animal.

The use of Freud's adjuvant is recommended. This is a water-in-oil emulsion which alone is called incomplete Freud's adjuvant (IFA), and incorporating inactivated or dead *Mycobacterium tuberculosis* bacteria is called complete Freud's adjuvant (CFA). It is an excellent adjuvant for stimulating a strong and prolonged immune response. The complete adjuvant does, however, induce persistent and sometimes aggressive granulomas. For this reason, the first injection should be given in the presence of CFA, and booster injections in the presence of IFA.

Antibody can first be detected in the serum approximately 7 days after initial injection, and persists for a few days. This primary response is usually weak. Booster injections, which may be adminsistered up to a year after the initial priming injection, result in a stronger response, peaking at around days 10–14 after injection, and persisting for up to a month. The response to third and subsequent injections is 'hyperimmunization': higher titres of antibodies are achieved, there is a class shift from IgM antibodies to predominantly IgG antibodies, and the response persists for longer.

Small serum samples should be taken 7–14 days after each immunization and tested for anti-lectin antibody titres. An acceptable level of antibody is usually detectable after the third and subsequent injections. At this point, larger volumes of 10–50 ml blood may be harvested.

2.3.2 Testing the antisera for anti-lectin activity

The most appropriate method of testing antisera that is required for immunohistochemistry, is to use an immunohistochemical technique. Ideally, tissue sections or preparations known to be strongly positive for the binding of a particular lectin should be used. The sections should be stained, using any of the methods described in Chapter 4, and incorporating the antisera at a range of dilutions. Doubling dilutions ranging from about 1/5 to about 1/100 are suggested. A good titre would be expected after about the second or third booster injection. An alternative, quicker, cheaper and cruder estimation of activity may be obtained by dot blotting. A suitable method is given in *Box 6.5* in Section 6.3. Details of alternative methods of assessing antibody titre are given in the excellent book by Harlow and Lane (1988).

2.3.3 Toxicity of lectins

Although many lectins are harmless, some are highly toxic. Good examples are ricin (from *R. communis*, the castor oil bean) and mistletoe (*V. album*) lectin which would be fatal if injected untreated; many others have severe or unknown toxicity. If it is known or suspected that a lectin is toxic, it should be denatured before preparing it for immunization. This generally does not adversely affect immunogenicity. A method for denaturing toxic lectins is given in *Box 2.7*.

Box 2.7: Denaturation of toxic lectins prior to immunization

For lectins of moderate or unknown toxicity.
1. Prepare the lectin solution in saline at the required concentration in the presence of 0.1 M monosaccharide to preserve the binding site [e.g. in the case of *H. pomatia* lectin (HPA from the Roman snail) the solution would be prepared in the presence of 0.1 M galNAc, the monosaccharide with the greatest inhibitory potential for HPA].
2. Heat the lectin solution for 30 min at 60°C.

For highly toxic lectins (e.g. ricin, abrin, mistletoe).
1. Treat the lectin as described above, then
2. Denature it further in 4% (v/v) formaldehyde solution in water overnight.
3. Dialyse extensively against saline to remove formaldehyde (note 1) (see *Box 2.4*).

Note
1. It is obviously critical to dialyse out as much as possible of the formaldehyde prior to immunization, and in practice this requires very extensive dialysis against large

(*continued*)

volumes of saline. In the authors' experience, an effective way to overcome this problem is to buy the lectin immobilized on agarose or Sepharose beads (or immobilize the lectin on beads 'in house' – the method is given in *Box 2.2*). The lectin on beads is denatured in 4% (v/v) formaldehyde solution in water overnight, as described above. Removal of the formaldehyde is then very easy:

- allow the beads to settle and pour away excess formaldehyde;
- resuspend in saline, agitate gently to wash, allow the beads to settle, and pour off excess liquid;
- repeat three or four times.

The beads may be mixed with adjuvant as described in the protocol given earlier and injected straight into the animal. The presence of agarose or Sepharose beads does the animal no harm and does not compromise the quality of antisera obtained; in fact, lectins immobilized on beads are very effective immunogens even in the absence of adjuvant.

2.4 An alternative to using antibodies: labelled lectins

One way of avoiding the need to buy or make antisera against lectins is to use them in a directly labelled form. All but the most obscure lectins can be purchased from commercial suppliers ready conjugated to a wide range of convenient labels such as peroxidase, alkaline phosphatase, FITC and TRITC. In the authors' experience, biotinylated lectins, which may be used in conjunction with avidin or streptavidin labelled with enzyme or fluorescent labels, are particularly versatile.

If no lectin conjugate is available, it may prove necessary to label lectins in the laboratory. Some useful methods are given in *Boxes 2.8–2.12*. In the authors' experience, the method for biotinylation given in *Box 2.8* is particularly simple, cheap and foolproof, although all labelling methods given are easy to carry out and effective.

Box 2.8: Labelling lectins with biotin
(after Bayer and Wilchek, 1980; Bayer *et al.*, 1976; Guesdon *et al.*, 1979)

1. Dissolve the lectin in 0.1 M NaHCO$_3$ at a concentration of 1 mg ml^{-1}, or dialyse lectin solution overnight against 0.1 M NaHCO$_3$.
2. Remove N-hydroxysuccinimidobiotin from the deep freeze 30 min before it is required, and allow to thaw at room temperature (see note 1).
3. Dissolve the N-hydroxysuccinimidobiotin in dimethylformamide or dimethyl-sulphoxide at a concentration of 1 mg ml^{-1}.
4. Immediately, add the biotin solution to the lectin solution in a ratio of approximately 120 µl biotin:1 ml lectin, and mix well (see note 2).
5. React at room temperature, preferably with constant gentle mixing (e.g. in an end-over-end mixer), for 2 h or overnight.

(continued)

6. Dialyse against TBS or lectin buffer (see *Box 2.4*) containing 0.1% (w/v) sodium azide, overnight.
7. Store at 4°C.

Notes
1. This is an important step because it is critical to exclude water from the *N*-hydroxysuccinimidobiotin, and removing the top from a frozen vial will allow water vapour to condense on the contents.
2. Once the *N*-hydroxysuccinimidobiotin is in solution, add it to the lectin solutions as quickly as possible.

Box 2.9: Labelling lectins with horseradish peroxidase
(after Nakane and Kawaoi, 1974; Tijssen and Kurstak, 1984)

1. Dissolve 5 mg of horseradish peroxidase in 1.2 ml distilled water.
2. Prepare 0.1 M sodium periodate in 10 mM sodium phosphate, pH 7.
3. Add 0.3 ml of the sodium periodate mixture to the horseradish peroxidase, mix well, and allow to react at room temperature, preferably with constant gentle mixing (e.g. in an end-over-end mixer) for 20 min.
4. Dialyse the mixture against several changes of 1 mM sodium acetate, pH 4, at 4°C, overnight.
5. Dissolve 5 mg of lectin in 0.5 ml of 20 mM carbonate, pH 9.5.
6. Add the dialysed peroxidase preparation to the lectin solution, mix well, and allow to react at room temperature, preferably with constant gentle mixing (e.g. in an end-over-end mixer) for 2 h.
7. Prepare a 4 mg ml^{-1} solution of sodium borohydride in distilled water.
8. Add 100 µl sodium borohydride solution to the lectin–peroxidase mixture, mix well, and allow to react at 4°C, with constant gentle mixing for 2 h.
9. Dialyse against several changes of TBS or lectin buffer (see *Box 2.4*), overnight.
10. Store at 4°C.

Box 2.10: Labelling lectins with alkaline phosphatase
(after Avrameas, 1969; Avrameas and Ternynck, 1969)

1. Dissolve 10 mg of lectin and 5 mg of alkaline phosphatase in 1 ml 0.1 M sodium phosphate buffer, pH 6.8 (see note 1).
2. Dialyse the mixture against several changes of 0.1 M sodium phosphate buffer, pH 6.8, at 4°C, overnight. Alternatively, desalt using a desalting chromatography column with an exclusion limit of approximately 20–50 kDa (note 2).
3. In a fume hood, stir in 0.05 ml of 1% (v/v) glutaraldehyde solution (EM grade), continue stirring for 5 min.
4. Leave for 3 h, in the fume hood, at room temperature.
5. Stir in 0.1 ml of 1 M ethanolamine in distilled water, pH 7.
6. Leave for 2 h, in the fume hood, at room temperature.
7. Dialyse against several changes of TBS or lectin buffer (see *Box 2.4*), at 4°C, overnight. Alternatively, desalt using a desalting chromatography column with an exclusion limit of approximately 20–50 kDa.
8. Centrifuge at 40 000 g for 20 min.
9. Draw off the supernatant and discard any precipitate.
10. Add 0.02% sodium azide, and store at 4°C in the presence of 1 mM $MgCl_2$, 1 mM $ZnCl_2$ and 50% glycerol.

(*continued*)

Notes

1. Alkaline phosphatase is often supplied as a precipitate in 65% saturated ammonium sulphate. A convenient way of making up the lectin/alkaline phosphatase mixture is therefore to spin down the precipitate (bench-top centrifuge, 600 g, 10 min) and resuspend it in a 10 mg ml⁻¹ solution of the lectin in 0.1 M sodium phosphate buffer, pH 6.8.
2. This step is necessary to remove any free amino groups present in the alkaline phosphatase–ammonium sulphate precipitate.

Box 2.11: Labelling lectins with fluorescein or rhodamine isothiocyanate (after The and Feltkamp, 1970a,b; Goding,1976)

1. Dissolve the lectin at a concentration of 2 mg ml⁻¹ in 0.1 M sodium carbonate in distilled water, pH 9.
2. Dissolve FITC or TRITC at a concentration of 1 mg ml⁻¹ in dimethyl sulphoxide.
3. Add 50 µl of FITC or TRITC solution to 1 ml of lectin solution; do this slowly, adding the fluorescent dye in 5 µl drops. Stir gently and continuously.
4. Allow to react in the dark, with constant gentle agitation (e.g. in an end-over-end mixer) for 8 h or overnight, at 4°C.
5. Add enough NH₄Cl to give a final concentration of 50 mM, and incubate at 4°C for 2 h.
6. Add enough xylene cyanol to give a 0.1% (v/v) solution, and enough glycerol to give a 5% solution.
7. Either pass the mixture through a desalting column with an exclusion limit of 20–50 kDa (labelled lectin elutes first, while unincorporated label is retained by the column) or dialyse against several changes of TBS or lectin buffer (see Box 2.4), overnight.
8. Store at 4°C in a dark container in lectin buffer containing 0.01% sodium azide.

Box 2.12: Labelling lectins with FLUOS [5(6)-carboxyfluorescein-*N*-hydroxysuccinomide ester]

1. Dissolve 10 mg lectin in 1 ml 0.1 M NaHCO₃ in distilled water, pH 8.5.
2. Add a solution of 0.85 mg FLUOS (Boehringer Mannheim) in 850 µl dimethylsulphoxide to the lectin solution.
3. Shake the reaction mixture at room temperature for 1 h.
4. Separate the conjugate from unbound FLUOS by either passing the mixture through a column of Sephadex G50 and washing through with 0.1 M NaHCO₃ buffer (the lectin conjugate passes through the column unhindered, while free FLUOS is retarded) or dialyse against several changes of TBS or lectin buffer (see Box 2.4), overnight.

References

Avrameas S. (1969) Coupling of enzymes to proteins with glutaraldehyde. Use of the conjugates for the detection of antigens and antibodies. *Immunochemistry* **6**, 43–52.

Avrameas S, Ternynck T. (1969) The cross-linking of proteins with glutaraldehyde and its use for the preparation of immunoadsorbants. *Immunochemistry* **6**, 53–66.

Bayer EA, Wilchek M, Skutelsky E. (1976) Affinity cytochemistry: the localisation of lectin and antibody receptors on erythrocytes via the avidin–biotin complex. *FEBS Lett.* **68**, 240–244.

Bayer EA, Wilchek M. (1980) The use of avidin–biotin complex as a tool in molecular biology. *Meth. Biochem. Anal.* **26**, 1–45.

Goding JW. (1976) Conjugation of antibodies with fluorochromes: modifications to the standard methods. *J. Immunol. Methods* **13**, 215–226.

Guesdon J-L, Ternynck T, Avrameas S. (1979) The use of avidin–biotin interaction of immunoenzymatic techniques. *J. Histochem. Cytochem.* **27**, 1131–1139.

Harlow E, Lane D. (1988) *Antibodies: a Laboratory Manual.* Cold Spring Harbor Laboratory Press, Cold Spring Harbor, NY.

Nakane PK, Kawaoi A. (1974) Peroxidase-labelled antibody: a new method of conjugation. *J. Histochem. Cytochem.* **22**, 1084–1091.

Porath J, Kristiansen T. (1975) Biospecific affinity chromatography and related methods. In: *The Proteins,* Vol. I (eds H Neurath, RL Hill, C-L. Boeder). Academic Press, New York, pp. 95–178.

Pusztai S, Bardocz S, van Damme EJM, Peumans WJ. (1997) *Handbook of Plant Lectins: Properties and Biomedical Applications.* John Wiley & Sons, Chichester, in press.

The TH, Feltkamp TEW. (1970a) Conjugation of fluorescein isothiocyanate to antibodies. I. Experiments on the conditions of conjugation. *Immunology* **18**, 865–873.

The TH, Feltkamp TEW. (1970b) Conjugation of fluorescein isothiocyanate to antibodies. II. A reproducible method. *Immunology* **18**, 875–881.

Tijssen P, Kurstak E. (1984) Highly efficient and simple methods for the preparation of peroxidase and active peroxidase–antibody conjugates for enzyme immunoassays. *Analyt. Biochem.* **136**, 451–457.

3 Lectin Histochemistry for Light Microscopy: I. Principles, Sample Preparation and Choice of Label

3.1 Definition of lectin histochemistry

Lectin histochemistry may be defined as the binding of a lectin to tissue-bound carbohydrate residue, detected by means of a visible label.

3.2 A philosophical point before starting

Although numerous lectins have been commercially available for years, remarkably little has been published on the use of most of them. Histologists seem reluctant to try something new and the majority of lectin publications report on the same dozen or so lectins, frequently purchased as a lectin kit and chosen by the chemical company according to availability rather than on a scientific basis. So be daring. Good science is about asking the right question or defining an important problem and trying to solve it, not simply about the use of expensive equipment.

We recommend:

1. Choose a biological phenomenon, a diagnostic problem or some *a priori* hypothesis to test. Otherwise, select one lectin and explore

the staining of a range of tissues; or select one tissue and explore a selection of lectins with a theme, such as gal-specific or fuc-specific lectins.

2. Start by using a well characterized lectin that is almost certain to work.
3. Progress to lesser known lectins.
4. Look for new, uncharacterized lectins (see Section 2.4 on labelling lectins for histochemistry).

3.3 Basic requirements for successful lectin histochemistry

For technically successful lectin histochemistry, one needs three basic things:

1. A suitable lectin.
2. Preservation of carbohydrate structures in the cell or tissue preparation.
3. A visible label.

The range of lectins available, and their sources, have been covered in the previous chapters. The preservation of carbohydrate structures in the cell or tissue preparation depends upon choice and preparation of specimens, which is described in Section 3.4, and choice of label is covered in Section 3.5. Finally, the range of methods available for detection of lectin binding to cells or tissues is described in Chapter 4, along with some foolproof methods for the beginner.

3.4 Preparation of specimens: cells and tissues

The advantages and disadvantages of various different cell and tissue preparations are discussed below and summarized in *Table 3.1*.

3.4.1 Cell suspensions

Lectin binding to glycoconjugates (glycoproteins and glycolipids) expressed at the surface of living cells in suspension may be demonstrated very simply. Suitable sources of cells include: cells grown in tissue culture, suspensions of human or animal blood cells,

Table 3.1: Advantages and disadvantages of different cell and tissue preparations

Preparation	Suitable for	Advantages	Disadvantages
Cell suspensions	Living cells (e.g. blood cells, cultured cells, cells released from solid tissue masses, ascites, pleural effusions)	Unaltered glycoconjugate expression in the living cell seen. Excellent for cell surface glycoconjugates. Glycoproteins and glycolipids seen	Not suitable for demonstration of intracytoplasmic glycoconjugates
	Direct method using fluorescent-labelled lectins most suitable	Quick and easy	Ephemeral preparations – view and photograph immediately
Imprints	Any solid tissue of plant, animal or human origin	Very quick and easy. Many imprints made in a very short time	Morphology not as good as with other methods
	Any lectin staining method suitable	Good for cytoplasmic glycoconjugates Glycoproteins and glycolipids detectable	
Smears	Any living cells in suspension (e.g. blood cells, cultured cells, cells in ascites or pleural effusions)	Quick and easy Good for cytoplasmic glycoconjugates	Morphology sometimes indistinct
	Any lectin staining method is suitable	Glycoproteins and glycolipids detectable	
Frozen sections	Any fresh solid plant, animal or human tissue Any lectin staining method is suitable	Relatively quick Fairly good morphology Spatial relationships of cells within tissues seen Good for cytoplasmic and cell surface glycoconjugates Glycoproteins and glycolipids detectable	Technically more demanding than imprints, suspensions or smears

(continued)

Preparation	Suitable for	Advantages	Disadvantages
Paraffin sections	Any solid plant, animal or human tissue	Tissue preserved for prolonged period. Excellent morpholology	More time-consuming than other methods
	Any lectin-staining method suitable	Relationships between cells in tissues preserved	Glycolipids lost
		Both cell surface and cytoplasmic glycoconjugates can be detected	Fixation and processing may damage some glycoconjugates

or cells released from solid tissues and cancer cells in ascites or pleural effusions.

Step 1. Obtaining the cell suspension.

1. *Cultured cells.* Cells grown in tissue culture can be gently released from the surface of the tissue culture flask using a soft rubber scraper (NB. Try to damage the cells as little as possible). It is not necessary to have a single-cell suspension for lectin binding studies – lectin binding will work well on small- to medium-sized clumps of cells in suspension. Very large clumps of cells should be avoided as they will render microscopy difficult.

   It is preferable to use a mechanical method, such as gentle scraping, to release cells adhering to tissue culture plastic, rather than employ enzymic digestion (e.g. by trypsin or other proteases). Enzyme digestion may damage or strip glycoproteins of interest from the cell surface thus drastically altering the results of lectin binding studies. If, for some reason, enzyme digestion must be used, cells should be allowed sufficient time (i.e. at least 24–48 h) in culture, but with gentle stirring to prevent re-adhesion, to recover before lectin binding studies are carried out.

   An alternative to enzymatic release of adherent cells is the use of 2.2 mM EDTA in a 2% (w/v) solution of BSA in water. This will most likely release a single-cell suspension of cells without significantly damaging cell surface glycoconjugates.

2. *Blood cells.* Human and animal blood cells are very suitable for lectin binding studies but have a limited range of oligosaccharides expressed at their surface. Enzymic digestion of red blood cells (e.g. by neuraminidase) will reveal a wider range of subterminal oligosaccharides.

3. *Pleural effusions and ascites.* Cancer patients with widespread disease sometimes develop collections of fluid around their lungs (pleural effusion) or in their peritoneal cavity (ascites). Such collections of fluid may become very large and cause the patient discomfort. For this reason, the fluid is sometimes drained off by the patient's doctor. Pleural effusions and ascites from cancer patients are typically rich in suspended cancer cells, as illustrated in *Figure 3.1.*

4. *Cells released from solid tissues.* It is possible to release cells into suspension from a solid tissue mass. This works best using soft, homogeneous tissue such as liver, but may be attempted on any human or animal tissue. A small piece of fresh tissue is cut using a sharp scalpel or razor blade. The tissue is then gently mashed or teased using either the scalpel blade or some other suitable instrument such as a razor blade. This approach inevitably leads to considerable cell damage, and if lectin binding studies on solid tissues are to be attempted, a different method (such as frozen or paraffin sections) might be more appropriate. If, however, cells are to be released from a solid tissue mass, several washes and resuspensions in tissue culture fluid or lectin buffer should be carried out in an attempt to remove as much cellular debris as possible.

   As with cultured cells, it is far preferable to use a mechanical method of gentle mashing and teasing to release cells from the tissue mass than to use enzyme digestion (e.g. by trypsin or other proteases). Enzyme digestion may damage or strip glycoproteins of interest from the cell surface thus drastically altering the results of lectin binding studies. If, for some reason, enzyme digestion must be used, cells should be allowed sufficient time (i.e. at least 24–48 h) in culture to recover before lectin binding studies are carried out.

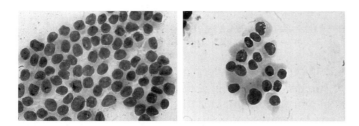

Figure 3.1:   Cancer cells in suspension in a pleural effusion.

Step 2. Suspending cells in lectin buffer. It is important to remember that any glycoconjugates in the fluid bathing the cells (e.g. tissue culture fluid or serum) may competitively inhibit lectin binding. For this reason, cells should be suspended in lectin buffer containing 2% (w/v) BSA prior to incubation with the lectin (the method is given in *Box 3.1*). The recipe for lectin buffer is given in Appendix A.

Box 3.1: Suspension of cells in 2% (w/v) BSA in lectin buffer

1. Place the cells (suspended in tissue culture fluid, serum, etc.) in a suitable centrifuge tube (e.g. a 50 ml plastic Falcon tube). Pellet the cells by gentle centrifugation in a bench-top centrifuge at approximately 200–300 *g* for 5 min.
2. Draw off and discard the supernatant.
3. Add lectin buffer with 2% (w/v) BSA, and resuspend the cells by very gently rocking the tube either by hand or in an end-over-end mixer. (NB. Too vigorous agitation will damage the cells. Do not shake the tube.)
4. Pellet the cells by gentle centrifugation as in no. 1.
5. Draw off and discard the supernatant.
6. Add lectin buffer with 2% (w/v) BSA, and resuspend cells again by gently rocking the tube as in no. 3.

Troubleshooting

Poor cell morphology is usually caused by mechanical damage to the cells. This can be minimized by treating cells as gently as possible. For example, when re-suspending cells, gently roll the tube back and forth, or use an end-over-end mixer. Do not shake or treat roughly. Similarly, when removing cells from tissue culture flasks, scrape very gently with a rubber scraper, using minimum pressure. Reduce the centrifugation steps: spin at a lower speed, for a shorter time.

Cell morphology may also be damaged through osmotic shock. Ensure that cells are at all times suspended either in body fluids, tissue culture fluid or carefully prepared iso-osmolar lectin buffer.

Use cells as quickly as possible after preparation. If they must be stored, store in serum-enriched tissue culture medium at 4°C.

Step 3. Lectin binding. The most appropriate method for lectin binding to cell suspensions is a direct method using a fluorescent label, such as FITC or TRITC. A suitable method is listed in *Box 4.1*, and examples of results illustrated in *Figure 3.5*.

3.4.2 Cell smears and cell imprints

Imprints. Imprints can be made from any solid tissue from plant, animal or human source. They are very quick and simple to prepare (a large number can be prepared in a few minutes) and can give moderately good morphological detail. The method for their preparation is given in *Box 3.2*.

Box 3.2: Preparation of imprints

1. Wear gloves. Cut a small piece of tissue, approximately 1 cm³, using a clean, sharp scalpel or razor blade.
2. Holding the tissue block firmly but gently using forceps, touch the freshly cut face of the tissue block briefly and lightly on to a clean glass microscope slide. (NB. Do not press or smear the tissue, simply touch it on to the glass quickly, cleanly, lightly.) Cells will become detached from the face of the tissue block and stick to the glass. The first few imprints may be too thick or bloody to use and should be discarded. Many (>20) good quality imprints can be made from the same cut surface.
3. Air dry the imprints at room temperature for 15 min or fix them by dipping briefly in 70% (v/v) ethanol in distilled water or cold (–20°C) 100% acetone (acetone will elute glycolipids), and then allow to air dry.

Imprints can then be stained immediately for lectin binding (any of the methods listed in Chapter 4 are suitable) or can be wrapped individually in foil and stored frozen at –20°C until required.

Troubleshooting
With some tissues, excellent morphological detail can be achieved with tissue imprints, others will never be very satisfactory. The individual worker must experiment to see what works well and what does not. Having said that, morphological detail is most often lost due to poor preparation of the imprints. The microscope slide must be very clean and dry – avoid smudgy fingerprints or greasy marks. It is a good idea to have a supply of microscope slides in a jar of 100% ethanol. Slides may be removed, quickly dried with a soft cloth, and used as required. An adhesive such as poly-L-lysine (see *Box 3.3*) may improve results. Most importantly, the freshly cut surface of the tissue should be briefly and cleanly touched on to the glass; do not press, drag, or smear the tissue.

Box 3.3: Coating slides with poly-L-lysine adhesive (see *Figure 3.2*)

1. Place a tiny drop of 0.01% (w/v) poly-L-lysine in distilled water on the centre of a clean microscope slide, approximately 0.5 cm from one end.
2. Place a second clean slide on top of the first, at right angles to it, so that the two slides form an 'L' shape, and the poly-L-lysine drop spreads between them.
3. Rotate the slides by 90° so that they lie in a straight line.
4. With a quick, clean, smooth action, drag one slide away from the other so that the poly-L-lysine very thinly coats both slides.
5. Allow slides to air dry for 5 min before use. Once coated and dried they may be stored in an enclosed dust-free box until required.

Notes
1. Poly-L-lysine has a strong positive charge, attracting the predominantly negatively charged cell/tissue to the slide.
2. Poly-L-lysine may be purchased as a concentrate or a ready to use solution. At its working concentration of 0.01% (w/v) it is stable for some weeks at 4°C. We usually aliquot either the working dilution or a 0.1% (w/v) concentrate and store frozen until required.

Smears. Smears can be made from any suitable suspension of cells, for example, blood, cultured cells in suspension, pleural effusions, ascites. The method for their preparation is given in *Box 3.4*. Smears

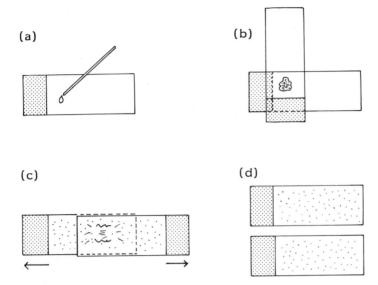

Figure 3.2: A method for coating slides with poly-L-lysine adhesive or for making cell smears.
(a) Place a drop of poly-L-lysine, or cells, in the centre of a microscope slide, approximately 0.5 cm from one end. (b) Place a second microscope slide on top of the first, at right angles to it, so that the drop spreads between them. (c) Rotate the slides so that they lie in a straight line, and at the same time drag one away from the other so that a film thinly coats both slides. (d) Allow the coated slides to dry, then use one or store until required.

may be stained at once for lectin binding (any of the methods listed in Chapter 4 are suitable) or can be wrapped individually in foil and stored frozen at –20°C (or colder) until required.

Box 3.4: Preparation of cell smears

The method for preparing cell smears is essentially the same as that for coating slides with the adhesive poly-L-lysine, as already described in *Box 3.3*. *Figure 3.2*, illustrating the technique for coating slides with poly-L-lysine, therefore illustrates the principle of this technique too.

1. Take a clean glass microscope slide and grasp it firmly at one end (if frosted microscope slides are used, at the frosted end). Place a drop of suspended cells in the central plane of the slide, approximately 5 mm from where it is being held.
2. Take a second clean glass slide and grasp it firmly at one end. Place it against the first slide, at right angles to it, so that the drop of cell spreads between the two slides.
3. Rotate the slides against each other through 90° so that they lie in a straight line.

(continued)

4. With a quick, clean, smooth action, drag them apart one against the other, so that the drop of cells is smeared equally along both of them. Alternatively the drop of cells in step 1 may be spread over the surface of the slide by dragging the edge of a second slide through the drop and over its surface, as illustrated in *Figure 3.3.*
5. Allow the smears to air dry for 15 min or fix them by dipping briefly in 70% (v/v) ethanol in distilled water or cold (–20°C) 100% acetone, and then allowing them to air dry.

(a)   (b)   (c)

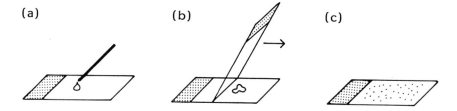

Figure 3.3:   An alternative method for making cell smears.
   (a) Place a drop of cell solution on a clean microscope slide. (b) Drag the edge of a second slide through the drop of cells, spreading a thin film. (c) Allow to air dry briefly. Use at once or store frozen, individually wrapped in foil, until required.

3.4.3   Frozen (or cryostat) sections

Frozen sections can be prepared from any suitable piece of fresh animal or human tissue by using a cryostat (basically a slicing machine contained within a chilled cabinet). Their preparation is more troublesome than that of smears or imprints, but the morphology is generally far superior. The method for the preparation of cryostat sections is given in *Box 3.5.*

Box 3.5: Preparation of cryostat sections

1. Cut a small piece of tissue, approximately 5 mm³ using a clean, sharp scalpel or razor blade. (NB. Use a very sharp scalpel blade; avoid crushing, tearing or otherwise damaging the tissue.)
2. Place the tissue on a cryostat chuck and enshroud it in a generous amount of 'optimum cutting temperature' sectioning medium (OCT; Raymond A. Lamb) or any suitable frozen sectioning medium.

(continued)

3. Snap freeze the tissue. This is best achieved by immersing chuck, OCT and tissue briefly in isopentane pre-cooled in liquid nitrogen. However, other methods are suitable, including spraying with a freezing spray (e.g. 'Cool Jet' from Raymond A. Lamb) or immersing briefly in liquid nitrogen.
4. Place the chuck and cooled tissue into the cabinet of the cryostat, and allow to equilibrate to cabinet temperature (approximately –20 to –30°C) for about half an hour.
5. Cut sections at 5–10 μm thick (generally speaking, it is easier to cut thicker sections, but the resolution will be poorer) and pick them up on clean, poly-L-lysine (see *Box 3.3*)–coated glass microscope slides.
6. Allow the sections to air dry for 15 min or fix them by dipping briefly in 70% (v/v) ethanol in distilled water or cold (–20°C) 100% acetone, and then air dry.

Sections may be stained at once for lectin binding (any of the methods listed in Chapter 4 are suitable) or can be wrapped individually in foil and stored frozen at –20°C (or colder) until required.

Warning: great care should be taken to avoid cutting oneself on the very sharp cryostat blade. This is especially important when working with unfixed tissues since possible pathogens (e.g. HIV, hepatitis B virus) in the tissue constitute a biological hazard.

Troubleshooting
1. Sections won't cut. There may be many reasons for this. The most common problem is that the temperature of the specimen, the chamber or the knife is not ideal. The correct temperature is critical for cutting good frozen sections! After snap freezing, the specimen should be left to equilibriate in the chamber for at least 30 min before cutting. When operating the cryostat, keep the chamber lid or door closed as much as possible so that warm air from the laboratory does not enter. Experiment with different temperature settings of chamber and knife (usually between –15°C and –30°C). Be patient! Sometimes it is a good idea simply to go away for a while and come back later to try again. Usually one finds that suddenly conditions just seem to come right and the sections start to cut well.
Apart from temperature, other factors that may affect section cutting include:
- Sharpness of the knife: if the knife is blunt or scored, this will adversely affect section cutting. Try moving the blade along a little and cutting in a different place.
- Cleanness of the knife: make sure that the knife is free of rust, grease or debris. During cutting, regularly clear tissue debris away from the cutting edge of the knife by gently brushing upwards (i.e. away from the cutting edge of the blade) with a soft paintbrush.
- The anti-roll plate: correct setting of the anti-roll plate is critical. If it is set incorrectly, refer to the manufacturer's instructions, and experiment until good quality sections are obtained.
- Size and shape of the tissue block: generally speaking a small (~5 mm × 5 mm) square face of tissue is easier to cut than a larger or eccentrically shaped chunk.
- Consistency of tissue block: soft homogeneous tissue like liver or kidney is generally much easier to cut than heterogeneous, fibrous or hard tissue. If you have little experience of cutting frozen sections, it is a good idea to practice on something easy (a piece of animal liver bought from a butcher's shop is ideal).

(*continued*)

Cutting good frozen sections takes skill and practice – be patient!

2. Sections float off microscope slide when stained. This is usually due either to the section itself being of poor quality (e.g. too thick, scored, incomplete or torn) or the microscope slide not being clean. Avoid fingerprints or greasy marks on the glass. Aim for top quality sections. Use of an adhesive like poly-L-lysine (see *Box 3.3*) will help.

3. Lectin binding is weak. This can sometimes be caused by soluble glycoconjugates (e.g. serum glycoconjugates) in the tissue section coming out into solution and competitively inhibiting lectin binding. In our hands, a brief (10 minute) wash in tap water, distilled water or lectin buffer prior to lectin binding can in some cases dramatically improve staining results.

3.4.4 Paraffin wax sections

These may be prepared from any suitable solid animal, human or plant tissue. There are many protocols for processing tissues to paraffin wax. The one given in *Box 3.6.* is a good general protocol for human/animal tissues. The authors have also enjoyed very successful results on assorted plant tissues (seeds, stems, roots, leaves) using the same methods. Individual workers may find that they prefer to adjust features of the protocol, or to try something different. A good reference guide is the book by Bancroft and Stevens (1990).

Box 3.6: Fixing and processing tissue for paraffin sections

Tissue cut by scalpel into small blocks. The easiest size to handle is approximately 1 cm × 1 cm × 5 mm. Blocks fixed in a suitable fixative for 12–48 h at room temperature or at 4°C.

A note about fixatives: a range of different fixatives are available (see Bancroft and Stevens, 1990, for details). The most commonly used are buffered formol saline or neutral buffered formalin; Bouin's fluid is also commonly used. Recipes are given in Appendix A. The tissue should be cut into chunks no bigger than 1 cm^3, fixed for 24 h, and then transferred to tap water or lectin buffer. The tissues may then be processed for embedding in paraffin wax. This can be done manually, but is most conveniently carried out in an automated tissue processing machine. A suitable protocol is listed below.

1. Cut the fixed tissue into suitably sized blocks, ideally approximately 5 mm × 5 mm × 2 mm.
2. Wash tissues in tap water overnight to remove the formalin.
3. Transfer to 70% (v/v) ethanol in distilled water for 24 h.
4. Pass through 2× changes of 70% (v/v) ethanol in distilled water for 1 h each.
5. Transfer to 80% (v/v) ethanol in distilled water for 2 h.
6. Transfer to 90% (v/v) ethanol in distilled water for 2 h.
7. Pass through 3× changes 100% ethanol for 2 h each.
8. Transfer to isopropyl alcohol for 2 h.

(continued)

9. Transfer to 1:1 isopropyl alcohol:chloroform for 2 h.
10. Pass through 2× changes of chloroform for 2 h each.
11. Pass through 2× changes of molten paraffin wax for 3 h each.

These solutions may be re-used two or three times before they require changing.

Once the tissue has been processed it may be embedded in paraffin wax blocks. The tissue should be surrounded on all sides by a border of paraffin wax at least 5 mm thick. The tissue block should be orientated squarely within the paraffin wax block to maximize ease of cutting.

Sections must be cut for lectin histochemistry to be carried out (see *Box 3.7*). They should be cut by microtome at 4–6 μm thick and may be used immediately or stored at room temperature for several years. Before use, sections must be dewaxed and rehydrated. The method for this is given in *Box 3.8*.

Box 3.7: Cutting paraffin sections by microtome

1. Cool the block on ice for 10–15 min.
2. Lock the block into position on the microtome, and advance it so that the surface of the block is fractionally below that of the knife.
3. Trim the block at 15 μm until sufficient wax has been trimmed away that the complete surface of the tissue block is exposed.
4. Cut sections of 4–6 μm thickness.
5. Float the sections out on the surface of a warm (approx. 40°C) water bath to flatten and stretch them.
6. Pick sections up on to clean glass slides (note 1), and allow them to drain briefly.
7. Dry the sections in a 60°C oven for 45 min, or on a hot plate at 60°C for 10 min.
8. Allow sections to cool and store them in a dust-tight container at room temperature.

Note
1. Poly-L-lysine coated slides (see *Box 3.3*) may be used.

Box 3.8: Dewaxing and rehydrating paraffin sections before use

1. Dewax the sections by soaking in xylene (see note 1) for 15 min.
2. Rehydrate the slides through two changes of 99% ethanol, 95% ethanol, 70% ethanol and finally distilled water. At each stage the slides should be agitated until they become equilibrated with their environment.

NB. Once dewaxed, sections must not at any point be allowed to dry out as this will result in an unacceptable level of non-specific background staining.

Notes
1. Xylene is an excellent solvent for paraffin wax but is hazardous and should be handled with care in a fume cupboard. Other solvents such as Histoclear may be used instead (Raymond A. Lamb) and are considered safer.
2. It is convenient to set up xylene, ethanols and distilled water in a series of Coplin jars (if a small number of slides are to be stained) or square glass dishes with suitable slide racks (if a larger number of slides are to be stained), as illustrated in *Figure 3.4*.

(continued)

Glassware and racks are available from Raymond A. Lamb and other suppliers. The dishes should be housed within a laboratory fume hood.
3. The solvents may be re-used several times before they need to be replaced.

All of the lectin staining methods detailed in Chapter 4 are suitable for paraffin sections.

Troubleshooting
1. Sections won't cut. There may be many reasons for this. Things to check include:
   - Is the block cold enough? Blocks should be chilled on ice for 15 min before cutting, and re-chilled by returning to ice or pressing an ice cube against the cut surface at intervals during cutting (i.e. every few minutes). This is especially important if the laboratory temperature is high, for example during the summer months.
   - Is the knife sharp enough? A blunt, damaged or scored knife will give poor results. Use of disposable blades, which should be changed regularly is recommended.
   - Is the knife clean? Make sure that there are no traces of rust or grease on the knife. During cutting, regularly clear tissue debris away from the blade by gently brushing upwards (i.e. away from the cutting edge of the knife) with a soft paintbrush.
   - The type of tissue. Soft, homogeneous tissue like liver and kidney is far easier to cut than heterogeneous, calcified, hard or fibrous tissue. If the tissue is very difficult to cut, you may simply need to be patient and persevere, or it may be a good idea to ask for assistance from an experienced member of the laboratory staff.
To cut good paraffin sections takes skill and practice – be patient!
2. Sections float off microscope slide when stained. This is usually due either to the section itself being of poor quality (e.g. too thick, scored, incomplete or torn) or the microscope slide not being clean. Avoid fingerprints or greasy marks on the glass. Aim for top quality sections. Use of an adhesive such as poly-L-lysine (see *Box 3.3*) will help.
3. Lectin staining is poor. It is important to optimize the staining method for any new lectin, on any new tissue. However, one special problem with paraffin sections is that the conditions of fixation and processing may damage and sequester glycoproteins, giving poor lectin staining or no staining at all. In many cases this can be addressed by use of a digestive enzyme such as trypsin or by microwave treatment. These options are fully explained later in this section (see *Boxes 3.9* and *3.10*).

Advantages of paraffin sections. The overwhelming advantage of using paraffin sections is that the preservation and morphology of the tissue is beautiful – far superior to that seen with even the finest frozen sections.

Paraffin-embedded blocks of tissue are small, neat and compact and may be maintained at room temperature for many years without deterioration: a very convenient way of storing specimens. Similarly, sections cut from paraffin-embedded blocks may be stored at room temperature for years until required for staining. Retrospective studies are thus possible.

Surgical specimens are routinely stored in the form of paraffin-embedded tissue blocks and paraffin sections. Most hospitals have archives of such tissues going back for many years or decades which,

Figure 3.4: Racks and glass dishes suitable for handling slides are available commercially and are very useful.

along with the medical records of the patients involved, provide invaluable material for clinical studies.

Potential problems with paraffin sections. If a quick result is required, paraffin sections are perhaps not the ideal choice, owing to the time and trouble involved in fixing and processing the tissue and preparing the sections. Frozen sections or imprints may be preferable. A more crucial problem is that during processing to paraffin, the impregnation with solvents such as chloroform washes out any lipids, including glycolipids, and these are irreversibly lost.

Generally speaking, carbohydrate structures are relatively tough and are undamaged by fixation and processing. However, when these carbohydrates form part of glycoprotein structures, the harsh conditions of fixation and processing may in some cases damage or sequester the protein, making the carbohydrate portion unavailable for lectin binding. Sometimes the damage is reversible, but sometimes it is not. If the damage is reversible, it may be repaired by treatment with a digestive enzyme (trypsin is most commonly used) or by microwaving. Methods are given in *Boxes 3.9* and *3.10*.

As a general rule, if paraffin sections are to be used it is a good idea to carry out lectin binding studies without the use of trypsin, and with trypsinization over a range of times (e.g. 5 min, 10 min, 20 min, 30 min) to ascertain which gives optimum results (a good background to

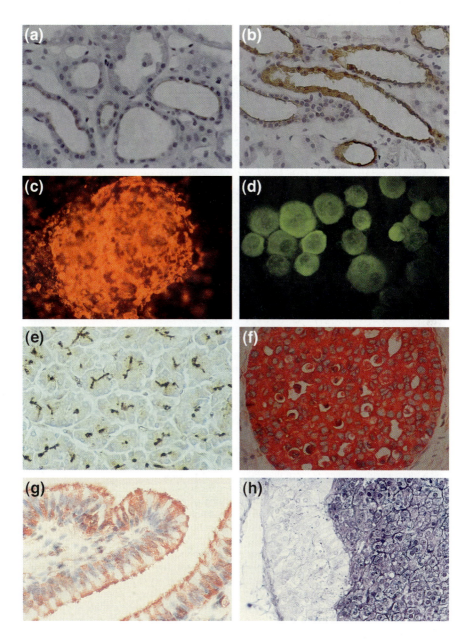

Figure 3.5: Trypsin digestion to reveal sequestered carbohydrates.
(a) PNA binding to normal human kidney (no trypsin). (b) PNA binding to normal human kidney (after 20 min trypsinization).
Examples of labels for lectin histochemistry.
(c) HPA binding to cultured human breast cancer cells labelled with TRITC. (d) DBA binding to cultured human colon cancer cells labelled with FITC. (e) HAA binding to pig pancreas labelled with horseradish peroxidase and DAB. (f) HPA binding to human breast cancer labelled with alkaline phosphatase and Fast Red. (g) HPA binding to human fallopian tube labelled with horseradish peroxidase and AEC. (h) Con A binding to rat adrenal cortex labelled with alkaline phosphatase and Fast Blue. Figure sponsored by Bradsure Biologicals and Dako Ltd.

target ratio). As a general rule, the length of trypsinization time required to give good results reflects the duration for which the tissue was fixed: long fixation time = long trypsinization time.

Trypsinization. This is a simple technique and can have quite dramatic effects in revealing carbohydrates sequestered by fixation and processing, as illustrated in *Figure 3.5*. In the authors' experience, when carrying out lectin binding experiments on paraffin sections, it is always well worth testing out the effect of a range of trypsinization times.

Box 3.9: Trypsinization of paraffin sections

1. Dewax the sections and rehydrate them through graded ethanol as described in *Box 3.8*.
2. Warm 400 ml of TBS or lectin buffer (pH 7.6; see Appendix A) warmed to 37°C in an incubator or waterbath. All glasswear etc. should also be warmed to 37°C.
3. Weigh out 400 mg trypsin (type II crude, from porcine pancreas; Sigma) and 400 mg calcium chloride and dissolve them in the warm buffer. Add the slides immediately, and incubate at 37°C (i.e. in the incubator or waterbath) for the required time (e.g. 5, 10, 20 or 30 min).
4. After incubation, remove the slides from the trypsin solution and wash in running tap water for approximately 5 min. They are then ready for lectin histochemistry.

Notes
1. Trypsinization should take place at 37°C. It is vital that the buffer and all glassware, etc., is pre-warmed, and that the incubation takes place in an appropriate waterbath or incubator.
2. The trypsin solution should be made fresh as required; the enzyme self-digests, and thus loses its activity over a period of approximately 30 min–1 h. For this reason, trypsinization times of more than 30 min are not usual.
3. Different tissues, fixed and processed in different ways, may require quite different trypsinization times. Thus it is worth trying a range of trypsinization times: too little trypsinization may result in sub-optimal lectin binding results; excessive trypsinization will lead to tissue degradation and loss of morphology and preservation.
4. The authors find that a crude trypsin preparation such as the one recommended above gives superior results to purer preparations. The reason for this is probably that the impurities (e.g. chymotrypsin) enhance the effect of the digestion.
5. Trypsinization is a fairly aggressive treatment – good quality sections are required.
6. Use of an adhesive such as poly-L-lysine (*Box 3.3*) is essential. incubate at 37°C (i.e. in the incubator or waterbath) for the required time (e.g. 5, 10, 20 or 30 min).

Box 3.10: Microwave treatment of paraffin sections (*Figure 3.6*)

1. Dewax the sections and rehydrate them through a series of graded ethanols as described in *Box 3.8*
2. Place the slides in coplin jars (no more than two sections per jar) and immerse them in citrate buffer (pH 6.0; see Appendix A for recipe).
3. Heat the slides in the microwave oven on full power until the buffer boils.
4. Allow them to simmer for the required time (eg. 2.5, 5 and 10 min).
5. Remove them from the microwave oven, and allow to cool to room temperature.

(continued)

6. Wash in several changes of tap water, distilled water or lectin buffer. The slides are then ready for lectin histochemistry.

Notes

1. In our hands, when the buffer reaches boiling point it tends to do so fairly explosively! Care should be taken. Watch what is happening. As the buffer nears boiling point, turn the power setting down to 'simmer' or 'medium'.

2. The buffer may spill due to explosive boiling (see above), and will certainly evaporate. Warm extra citrate buffer and distilled water in the microwave oven and top up the coplin jars as required. Slides should be well immersed at all times.

3. Do not add cold buffer or water to hot coplin jars (or vice versa) as this will cause them to crack or shatter.

4. This is a very aggressive treatment. Very good quality sections are required. Use of an adhesive such as poly-L-lysine (see *Box 3.3*) or something stronger such as Vectabond (Vector Laboratories – follow the manufacturer's instructions) is essential.

If poor results are obtained even when trypsin digestion is employed, it may be a good idea to experiment with the use of microwaving, again over a range of times such as 2.5 min, 5 min, 10 min.

3.5 Labels for detection of lectin binding

Many suitable labels exist for detection of lectin binding to cell or tissue preparations (see *Table 3.2*). They may be divided into two principal categories: fluorescent labels and enzyme (coloured) labels.

3.5.1 Fluorescent labels

Range of fluorescent labels available. Fluorescent labels fluoresce brightly when viewed under exciting ultraviolet or high energy blue light. Conventional fluorescent labels include FITC, which fluoresces green/yellow, and TRITC and Texas Red, which fluoresce red. Many lectins are commercially available ready conjugated to these labels and are extremely useful for direct labelling techniques (see Section 4.2). Examples of lectin binding labelled by TRITC and FITC are given in *Figure 3.5*. It is possible to obtain selected lectins conjugated to other fluorescent labels, from some commercial sources. It is also relatively simple to conjugate lectins to fluorescent tags (suitable methods are given in *Boxes 2.11* and *2.12*).

There are other fluorescent labels for use in histochemistry in general. It is possible to purchase secondary antibodies for 'sandwich' type methods (see Sections 4.3 and 4.5, for example) labelled with a range of fluorescent tags.

Figure 3.6: Carbohydrate retrieval may be carried out using a domestic microwave oven.

Uses of fluorescent labels. Fluorescent labels can give extremely elegant results. They can be used in conjunction with almost all of the staining methods listed in Chapter 4 and on virtually any cell or tissue preparation. However, they are most suitable for direct staining techniques, and in particular on cell suspensions, imprints or smears.

Limitations of fluorescent labels. Fluorescent labels do have drawbacks:

1. The fluorescence is ephemeral. Lectin histochemistry performed in conjunction with a fluorescent label should be viewed as soon as possible after completion because the fluorescence will fade with time (if, for some reason, sections cannot be viewed at once they should be stored at 4°C in the dark. The fluorescent signal will remain strong for a few hours, in some cases, a few days or even weeks). If a permanent record is required, the slides should be photographed.

2. Because cells and cell structures labelled by a fluorescent tag shine out against a black background like stars against a midnight sky, it is sometimes difficult to identify the exact cell type or structure being labelled. Non-labelled cells remain invisible. This can sometimes make interpretation of results troublesome. If problems are encountered, it is a good idea to switch quickly between fluorescence and phase contrast, dark field, Nomarski or ordinary bright field illumination. Counter-staining with a DNA stain such as DAPI might also help with the interpretation, although a DAPI filter set is not often supplied

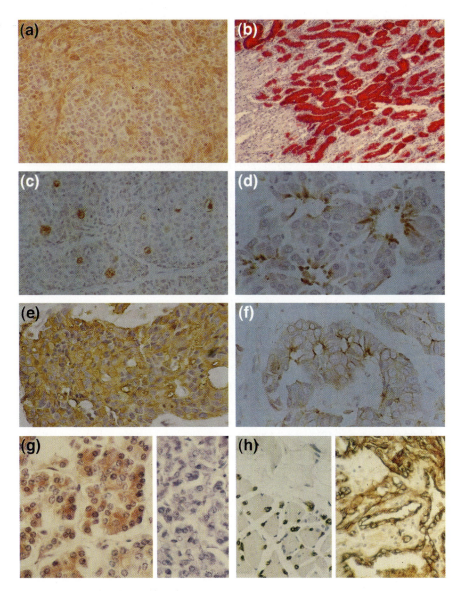

Figure 3.7: Problems with endogenous enzyme.

(a) Endogenous peroxidase in normal human spleen. (b) Endogenous alkaline phosphatase in normal human kidney.

Lectins with the same nominal monosaccharide binding specificity give very different binding patterns.

(c) VVA binding to breast cancer, (d) HAA binding to breast cancer, (e) HPA binding to breast cancer, (f) MPA binding to breast cancer.

The effect of neuraminidase digestion to strip terminal sialic acid.

(g) LFA binding to human pancreas, before and after neuraminidase digestion.

UEA-1 is an excellent marker of human epithelial cells.

(h) UEA-I binding to endothelium in rat striated muscle and to an endothelial tumour. Figure sponsored by Millipore (UK) Ltd, Biometra Ltd and Bio-Rad Laboratories Ltd.

with basic fluorescent microscopes. If a photographic record is required, it is often useful to take a photograph of fluorescent staining in parallel with the same field stained simply with haematoxylin and eosin (see *Box 3.21* for method) to aid interpretation.

3.5.2 Enzyme (coloured) labels

Range of labels available. The most commonly used enzyme labels are horseradish peroxidase and alkaline phosphatase. Many lectins and secondary antibodies, are commercially available ready labelled with one of these enzymes. It is also possible to label lectins or antibodies with them (methods for this are given in *Boxes 2.9* and *2.10*). Other enzyme labels (e.g. glucose oxidase or β-galactosidase) are less commonly used.

Enzyme labels are detected through their reaction with a colourless chromogenic substrate to yield a coloured product. A range of chromogenic substrates is available for each enzyme. Some useful methods are given in *Boxes 3.11–3.17* below. Examples of lectin staining using horseradish peroxidase as a label with DAB and AEC as the chromogenic substrates, and alkaline phosphatase as a label using naphthol-AS-BI-phosphate and Fast Red and Fast Blue as the chromogenic substrate are given in *Figure 3.5*.

Box 3.11: Chromogenic substrate for use with horseradish peroxidase – diaminobenzidine (DAB)

The most commonly used substrate for peroxidase. Very sensitive. Yields an *alcohol-insoluble* granular, deep brown product.

1. Prepare a solution of DAB tetrahydrochloride at a concentration of 0.5 mg ml^{-1} in TBS (pH 7.6) or in lectin buffer (see notes 1 and 2).
2. Add hydrogen peroxide to give a final concentration of 0.03% (v/v).
3. Apply to the specimen and incubate for 10 min (check colour development under the microscope).
4. Wash the specimen under running tap water for approximately 5 min.

The specimens are now ready to counterstain (*Box 3.20*), dehydrate, if a resinous mountant is to be used (*Box 3.22*), and mount (*Box 3.23*) for viewing and storage.

Notes
1. DAB is potentially carcinogenic and should thus be handled with great care. Follow local health and safety procedures for handling and disposal.
2. To minimize the hazards involved in weighing out DAB tetrahydrochloride powder, we find it convenient to buy the reagent in 1 g quantities. The entire 1 g of powder is dissolved in 200 ml distilled water, then 1 ml aliquots are quickly transferred to 200 × 10 ml stoppered, plastic tubes. The aliquots are stored frozen at −20°C until required. When required, a single aliquot is defrosted, 9 ml of buffer and 10 µl of 30% hydrogen peroxide solution added. It is then ready for use.

(continued)

3. Many companies market DAB in tablet form (e.g. Sigma) or dropper bottle kits (e.g. Vector Laboratories). These tend to work out slightly more expensive than simply purchasing the chemical in powder form, but are simple and safer to use.

4. Hydrogen peroxide solutions deteriorate: store a concentrated (30 vol.) stock solution at 4°C, in the dark. You can test whether the enzyme reaction works or not by adding a few drops of horseradish peroxidase-labelled lectin or antibody to the leftover DAB–hydrogen peroxide solution after you have finished using it. Formation of a dense granular brown precipitate confirms that the reaction is working well.

Box 3.12: Chromogenic substrate for use with horseradish peroxidase – aminoethyl carbazole (AEC)

AEC yields a granular, pink-red, *alcohol–soluble* product.

1. Dissolve 4 mg of AEC in 1 ml of *N,N*-dimethyl formamide (DMF) (see note 1).
2. Add 1 ml of the AEC solution to 15 ml of 0.1M sodium acetate buffer, pH 5.2 (for recipe see Appendix A).
3. Add hydrogen peroxide to give a concentration of 0.03% (v/v).
4. Filter the solution.
5. Apply to the specimen and incubate for 10–40 min (check colour development under the microscope).
6. Wash in running tap water for approximately 5 min.

The specimens are now ready to counterstain (*Box 3.20*) and mount in *aqueous* mountant (*Box 3.23*).

Note
1. A stock solution of AEC in DMF can be prepared in advance and stored at room temperature until required.

Box 3.13: Chromogenic substrate for use with horseradish peroxidase – chloronaphthol

Chloronaphthol yields a blue-black *alcohol-soluble* product.

1. Add 100 μl of a 3% solution of chloronaphthol (see note 1) in absolute ethanol to 10 ml of 0.05 M Tris, pH 7.6.
2. Filter the solution.
3. Add hydrogen peroxide to give a concentration of 0.03% (v/v).
4. Incubate with the specimen for 10–40 min until the colour develops (check colour development under the microscope).
5. Wash in running tap water for approximately 5 min.

The specimens are now ready to counterstain (*Box 3.20*) and mount using an *aqueous* mountant (*Box 3.23*).

Note
1. A stock solution of chloronaphthol can be prepared in advance and stored at −20°C.

Box 3.14: Chromogenic substrate for use with alkaline phosphatase – bromochloroindolylphosphate-nitroblue tetrazolium (NBT)

Probably the most sensitive label for use with alkaline phosphatase. Gives an intense purplish-black *alcohol-insoluble* product.

1. Prepare stock solution A: 0.5 g NBT in 10 ml 70% (v/v) dimethylformamide in distilled water (see note 1).
2. Prepare stock solution B: 0.5 g disodium bromochloroindolyl phosphate in 10 ml dimethylformamide (see note 1).
3. Prepare alkaline phosphatase buffer: 100 mM sodium chloride, 5 mM magnesium chloride, 100 mM Tris, pH 9.5 (see note 1).
4. Add 66 µl of stock solution A to 10 ml of alkaline phophatase buffer and mix.
5. Add 33 µl of stock solution B to the mixture and mix (see note 2).
6. Place slides in a suitable container (e.g. a Coplin jar or a rack inside a pot) and add substrate solution to cover. Incubate for approximately 30 min until colour develops (check progression of colour development under the microscope).
7. To stop the reaction, rinse in PBS containing 20 mM EDTA.
8. Wash in running tap water for approximately 5 min.

The specimens are now ready to counterstain (*Box 3.20*), dehydrate, if a resinous mountant is to be used (*Box 3.22*), and mount (*Box 3.23*).

Notes
1. Stock solutions A and B are stable at room temperature for at least 1 year.

Box 3.15: Chromogenic substrate for use with alkaline phosphatase – naphthol-AS-Bl-phosphate/Fast Red

Gives a bright red, *alcohol-soluble* product.

1. Prepare solution A: 20 mg naphthol-AS-Bl-phosphate in 200 µl dimethyl formamide or dimethyl sulphate.
2. Prepare solution B: 40 mg Fast Red in alkaline phosphate buffer (see previous section for recipe).
3. Add solution A to 100 ml alkaline phosphatase buffer and mix. Then add solution B and mix (see note 1).
4. Add 100 µl 1 mM magnesium chloride solution in distilled water, mix (see note 2).
5. Filter.
6. Place slides in a suitable container (e.g. a Coplin jar or a rack inside a pot) and add substrate solution to cover. Incubate for approximately 30 min until colour develops.
7. To stop the reaction, wash in running tap water for approximately 5 min.

The specimens are now ready to counterstain (*Box 3.20*), (check colour development under the microscope).

Notes
1. It is important to add solution A then B, in that order.
2. Use as soon as possible.

Box 3.16: Chromogenic substrate for use with alkaline phosphatase – naphthol-AS-BI-phosphate/New Fuchsin

Produces a bright pink-red *alcohol-insoluble* product.

1. Prepare solution A: 1 mg New Fuchsin in 0.25 ml of 2 M HCl.
2. Prepare solution B: 1 mg of sodium nitrate in 0.25 ml distilled water.
3. Prepare solution C: 10 mg naphthol-AS-BI-phosphate in 200 µl dimethylformamide or dimethyl sulphoxide.
4. Mix solution A and solution B. Shake for 1 min.
5. Add the mixture to 40 ml of 0.2 M Tris, pH 9.
6. Add solution C to the mixture.
7. Place slides in a suitable container (e.g. a Coplin jar or a rack inside a pot) and add substrate solution to cover. Incubate for approximately 30 min (check colour development under the microscope).
8. To stop the reaction, rinse in PBS containing 20 mM EDTA.
9. Wash in running tap water for approximately 5 min.

The specimens are now ready to counterstain (*Box 3.20*), dehydrate if a resinous mountant is to be used (*Box 3.22*) and mount (*Box 3.23*).

Box 3.17: Chromogenic substrate for use with β-galactosidase – 5-bromo-4-chloro-3-indolyl-β-D-galactopyranoside

β-galactosidase is not commonly used in histochemistry. It is nevertheless an excellent label. A good substrate is 5-bromo-4-chloro-3-indolyl-β-D-galactopyranoside which gives a deep blue *alcohol-insoluble* product.

1. Dissolve 4.9 mg of 5-bromo-4-chloro-3-indolyl-β-D-galactopyranoside in 100 µl dimethylformamide or dimethyl sulphoxide.
2. Add to 10 ml of PBS (for recipe see Appendix A) containing 1 mM magnesium chloride and 3 mM potassium ferrocyanide.
3. Filter.
4. Place slides in a suitable container (e.g. a Coplin jar or a rack inside a pot) and add substrate solution to cover. Incubate for approximately 30 min until colour develops (check progression under the microscope).
5. Stop the reaction by washing with running tap water for approximately 5 min.

The specimens are now ready to counterstain (*Box 3.20*), dehydrate if a resinous mountant is to be used (*Box 3.22*) and mount (*Box 3.23*).

Table 3.2: Comparison of labels for detection of lectin binding

Label	Chromogenic substrate	Appearance	Advantages	Disadvantages
Fluorescent labels				
FITC	—	Yellow-green	Excellent for direct labelling methods, especially for cell suspensions	Ephemeral. Photograph at once for a permanent record
TRITC	—	Red		
Texas Red		Red		
			Dramatic, photogenic results	Labelled structures fluoresce
			Very sensitive	against a black background sometimes making interpretation and identi-fication of cell type/tissue topography difficult
			Double labelling is possible	
Enzyme labels				
Horseradish peroxidase	DAB	Deep brown, granular	Alcohol-insoluble	DAB is potentially carcinogenic
			Good contrast for photography	?Unattractive colour
			Good morphology	
	AEC	Pink-red granular	Attractive colour – dramatic	Alcohol-soluble
	Chloronaphthol	Blue-black	Good contrast for photography	?Unattractive colour
				Alcohol soluble
Alkaline phosphatase	NBT	Purplish-black	Alcohol-insoluble	?Unattractive colour
			Good contrast for photography	
			Highly sensitive	
			Good for tissues high in endogenous peroxidase	
	Fast Red	Red	Attractive colour – dramatic	Alcohol-soluble
			Good for tissues high in endogenous peroxidase	

(*continued*)

Table 3.2 continued

Label	Chromogenic substrate	Appearance	Advantages	Disadvantages
	New Fuchsin	Pink	Attractive colour Alcohol-insoluble Good for tissues high in endogenous peroxidase	
Galactosidase	5-bromo-4- chloro-3- indolyl-β-D- galactopyranoside	Deep blue	Alcohol-insoluble Attractive colour Good contrast for photography	Not commonly used

Uses of enzyme labels. Enzyme labels may be used on virtually any tissue or cell preparation and in conjunction with any of the methods listed in Chapter 4.

Limitations of enzyme labels: the problem of endogenous enzyme. Many tissues contain appreciable amounts of endogenous peroxidase and/or alkaline phosphatase. Potentially, this can cause problems in staining and interpretation as endogenous enzyme will give the same colour reaction with the chromogenic substrate as the enzyme label (see *Figure 3.7* for some dramatic examples). This has implications when choosing a label for lectin histochemistry on a particular tissue type; for example normal spleen is extremely rich in peroxidase, therefore another label (either a fluorescent label or an alternative enzyme label like alkaline phosphatase) would be more appropriate. Similarly, kidney is exceedingly rich in alkaline phosphatase; when performing lectin histochemistry on kidney tissue, a label other than alkaline phosphatase may be appropriate (either a fluorescent label or an alternative enzyme label like horseradish peroxidase).

The potential problems of endogenous tissue enzymes are not insurmountable. Endogenous peroxidase may be quenched by treatment with hydrogen peroxide in methanol (the method is given in *Box 3.18*). Fixation and processing of tissues for embedding in paraffin wax usually destroys most or all endogenous peroxidase activity making it much less of a problem, although we would recommend that the blocking method described in *Box 3.18* is still incorporated in the staining protocol. Endogenous alkaline phosphatase may be inactivated by treatment with levamisole (the method is given in *Box 3.19*). This method is not suitable for intestinal tissues, as levamisole will not destroy intestinal alkaline phosphatase activity. If intestinal tissues are to be examined, choose another label such as peroxidase or a fluorescent label. Fixation and paraffin wax embedding will generally destroy alkaline phosphatase activity, and it should therefore not be a problem when staining paraffin sections.

Box 3.18: Blocking endogenous peroxidase

Immediately prior to incubation of cell or tissue preparation with the lectin.

1.   Immerse slides in a 1% (v/v) solution of hydrogen peroxide in methanol for 20 min.
2.   Wash in running tap water for approximately 5 min.

Sections are then ready for incubation with the lectin.

Notes
Once the methanol–hydrogen peroxide solution has been made up, it will remain active for approximately 3 days (less in hot weather, or if the laboratory is particularly hot) and can thus be re-used. It should be stored in a covered dish.

Box 3.19: Blocking endogenous alkaline phosphatase

Levamisole (0.1 mM) is incorporated in the substrate solution. Levamisole does not diminish the activity of the intestinal alkaline phosphatase used to label lectins and antibodies, but inhibits the activity of other tissue alkaline phosphatases. For this reason, when investigating lectin binding to intestinal tissue, a label other than alkaline phosphatase (e.g. horseradish peroxidase) may be more appropriate.

Counterstaining in conjunction with enzyme labels. When using an enzyme (coloured) label, a counterstain makes the interpretation of results easier. The counterstain generally labels cell nuclei.

Choice of counterstain depends on the colour of the staining product, for example using DAB–hydrogen peroxide in conjunction with a horseradish peroxidase label yields a brown coloured product: a blue haematoxylin counterstain contrasts with the brown and is highly suitable, giving an attractive and photogenic result. If a blue coloured product is obtained (e.g. when using 5-bromo-4-chloro-3-indolyl-β-D-galactopyranoside with a galactosidase label) clearly, haematoxylin is unsuitable, and an alternative counterstain should be chosen (see for example Bancroft and Stephens, 1990).

Many counterstains are available, but the most commonly used is haematoxylin. A method for its use is given in *Box 3.20* (see also *Box 3.21*).

Box 3.20: Use of haematoxylin counterstain

There are many different types of haematoxylin, but in the authors' experience the most appropriate for lectin histochemistry is Mayer's haematoxylin. This can be purchased ready made as a deep port wine-coloured liquid. When lectin staining is complete, slides are immersed in Mayer's haematoxylin for 2–3 min. They are then transferred to running tap water (see Section 4.1.2) where the slight alkalinity of the water turns the red stain deep blue. This process, for obvious reasons, is known as 'blueing' in tap water. If the tap water is not sufficiently alkaline, the nuclei will still appear purplish-pink. This may be remedied by either 'blueing' in tap water for longer, or 'blueing' in tap water to which a drop of ammonia has been added. The haematoxylin stains the nuclei of the cells.

Box 3.21: Quick haematoxylin and eosin stain

A excellent stain for revealing the general morphology of cell and tissue preparations. Nuclei stain blue, cytoplasm pink-red.

1. Prepare sections, imprints or smears as described in Sections 3.4.2, 3.4.3 and 3.4.4.
2. Immerse in Mayer's haematoxylin for 2–3 min.
3. 'Blue' in running tap water (see Section 4.1.2) for 5 min.
4. Immerse in 1% (w/v) eosin in water, 10 sec.
5. Rinse rapidly in tap water.
6. Dehydrate through graded alcohols, clear and mount (*Boxes 3.22* and *3.23*).

Mounting slides stained using enzyme labels. Some chromogenic substrates yield a coloured product which is insoluble in alcohol; some are alcohol soluble. This has implications for mounting the slides for viewing and/or storage. If an alcohol-soluble product is obtained, slides must be mounted in an aqueous mountant such as Aquamount (available commercially; e.g. from Raymond A. Lamb) or glycerol gelatin (Sigma). Generally speaking, aqueous mountants give poorer resolution than resinous mountants, and do not set to give a dry, permanent preparation. If an alcohol-insoluble product is obtained, slides can be mounted in a resinous mountant. The most commonly used resinous mountant is DPX, which is available commercially (Raymond A. Lamb). Prior to mounting in resinous mountant, slides must be dehydrated and cleared (the method for this is given in *Box 3.22*).

Box 3.22: Dehydrating and clearing slides for mounting in resinous mountant

1. Agitate slides for approximately 1 min in 75% (v/v) ethanol in water.
2. Transfer the slides to 95% (v/v) ethanol in water and agitate them for 1 min.
3. Transfer the slides through two changes of 99% (v/v) ethanol in distilled water and agitate them for 1 min in each.
4. Transfer the slides through two changes of xylene and agitate them for 1 min in each.

Slides are now ready for mounting in resinous mountant.

Notes
1. If a large number of slides are to be mounted, it is most convenient to place them in a rack and pass the rack through pots containing the solvents. If a smaller number of slides are to be mounted, coplin jars of solvents are appropriate.
2. At each stage, it is important that the slides are equilibrated with the solvent. Agitation (i.e. 'sloshing the slides up and down' in the solvent) speeds up this process.
3. The solvents may be re-used several times, but they will quickly become cross-contaminated and fresh solvents will be required. Carry-over of water into the clearing solvent will lead to tiny bubbles forming on the specimen.
4. If the clearing solvent becomes milky-white when the specimens are added, this indicates contamination of the solvent with water. Specimens should be returned, through the graded alcohols, to water. Fresh solvents should then be set up and the dehydration/clearing process repeated.
5. Slides may be left in the xylene overnight without deterioration.
6. Slides should not be allowed to dry out at any time.

Whether slides are to be mounted in aqueous or resinous mountant, the procedure is the same. This is given in *Box 3.23* and illustrated in *Figure 3.8.*

Box 3.23: Mounting slides (*Figure 3.8*)

1. Apply a drop of mountant to the centre of a glass coverslip using the tip of a Pasteur pipette or a wooden applicator stick.
2. Gently lower the glass slide, specimen facing downwards, on to the coverslip. Avoid air bubbles.
3. Carefully position the coverslip, pressing to spread the mountant and to expel any air bubbles.
4. Wipe away any excess mountant using a clean tissue.

Notes
1. Slides mounted in either aqueous or resinous mountant may be examined immediately. Resinous mountant will harden overnight to give a permanent preparation. Aqueous mountants usually remain sticky, but slides can be stored for weeks or months as long as they are kept separate and flat.
2. Upon storage, after mounting in resinous mountant, if bubbles appear in the mounted preparation, this suggests that the solvents were contaminated with water. Slides may be de-coverslipped and re-mounted, using fresh solvents, as follows. De-coverslip the slides by soaking in xylene overnight (slides that have been mounted for some weeks or months may require soaking for longer, perhaps 2 or 3 days) until the coverslip falls off. The process can be speeded up once the mountant has softened by gently easing the coverslip off using forceps, although this must be carried out with care to avoid damage to the tissue section. They can then be remounted by first rehydrating through graded ethanols to water (*Box 3.8*) then dehydrating through graded ethanols and clearing in xylene (*Box 3.22*) and re-mounting (*Box 3.23*).

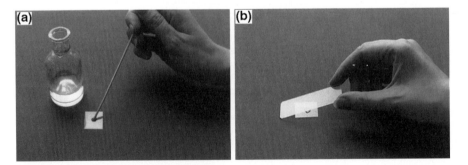

Figure 3.8:   Method for cover slipping.
(a) Place a drop of mountant on the coverslip. (b) Lower the microscope slide into place, being careful to exclude air bubbles.

Reference

Bancroft JD, Stevens A. (1990) *Theory and Practice of Histological Techniques,* 3rd Edn. Churchill-Livingstone, Edinburgh.

4 Lectin Histochemistry for Light Microscopy: II. Methods for Visualization of Lectin Binding

The basic methods for detection of lectin binding to cells or tissues are given in this chapter. There are, however, many variations on these basic techniques, all of which work equally well, and the individual worker is encouraged to experiment.

4.1 General practical notes

4.1.1 Incubation in a humid chamber

In the methods that follow, it is recommended that all incubations with lectins, antibodies, etc., are carried out in a humid chamber. The presence of a humid atmosphere prevents evaporation, keeping the concentration of the lectin, antibody, etc., stable. It is possible to purchase specially designed (usually perspex) chambers for this purpose, or to instruct 'in house' workshops to manufacture them. A typical design is given in *Figure 4.1*. A few millimetres of water is placed in the bottom of the chamber before staining, and the lid kept in place during incubation periods. If large-scale histochemistry is planned, these are a very good investment.

It is, however, possible to set up a basic chamber quite simply. Small-scale histochemistry can be carried out in a series of Petri dishes lined with discs of wet blotting paper. The slides may be simply laid on top of the paper and the lid fitted into place (*Figure 4.2*).

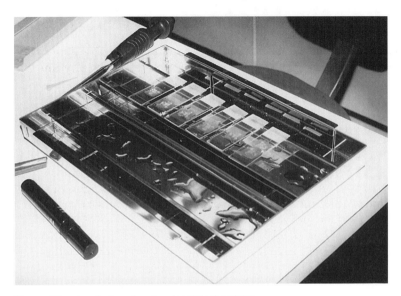

Figure 4.1: Incubation chamber for histochemistry.

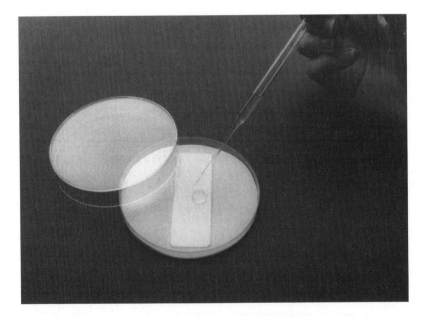

Figure 4.2: Use of a Petri dish lined with damp blotting paper as a makeshift incubaton chamber.

Alternatively, an effective chamber can be created by arranging glass rods embedded in plasticine in the bottom of a plastic sandwich box. As with the commercially purchased chamber, a few millimetres of water is placed in the bottom of the box to maintain humidity, and the lid kept in place during incubation periods.

4.1.2 Washing in running tap water

At various stages in staining protocols, washing slides in running tap water is recommended. If a few slides only are being stained, they should be stood upright in a glass Coplin jar. If a larger number of slides are being stained, plastic or metal racks (typically holding 12 or 25 slides) may be purchased from suppliers such as Raymond A. Lamb. Slides can be placed in the rack and the rack placed inside a sandwich box or glass dish. A rubber tube fitted to the laboratory tap can then be directed into the Coplin jar or dish, and a gentle stream of water directed into it, as illustrated in *Figure 4.3*. (NB. The flow of water should not be so powerful as to damage cell or tissue preparations.)

4.1.3 Washing in lectin buffer

Thorough washing is essential for good staining results. The most effective method of washing is to place the slides either upright in a

Figure 4.3:   Washing slides in running tap water.

Coplin jar of buffer, or, if a larger number of slides are being stained, in a slide rack immersed in a container of buffer. Slides should be agitated vigorously by 'sloshing up and down' in the buffer for a few seconds, and then allowed to stand in buffer for 1–2 min. At least three washes, each time in fresh buffer, are necessary to ensure clean staining results.

4.1.4 Drying out

It is absolutely essential that slides should not be allowed to dry out at any time during the staining procedure as this will lead to dirty, non-specific background staining and poor, unreliable results.

4.1.5 Incubation with lectin or antibody

The recommended method for incubating slides with lectin or antibody is as follows:

1. Allow approximately 100 µl of antibody or lectin solution per slide.
2. Remove slide from washing buffer, water, etc., and quickly shake off any excess.
3. Wipe the back of the slide dry using a clean tissue.
4. Very carefully, wipe around the cell or tissue preparation to remove excess moisture. Be very careful not to touch or damage the preparation itself. Do not let the preparation dry out at any time.
5. Place the slide, face up, in the humid chamber.
6. Using a Pasteur pipette, drip on enough lectin or antibody solution to just cover the cell or tissue preparation.
7. Place the lid on the chamber and leave for the recommended incubation time.

4.1.6 Optimum temperature for incubation

It is usual to carry out most staining protocols at normal room temperature (in almost all cases, this should give good results), however it is sometimes worth experimenting.

As a general rule with antibodies, staining intensity will be increased, or shorter incubation periods are required, if incubations are carried out at 37°C instead of room temperature. Binding will usually be weaker, or take longer, at 4°C (for example, it is sometimes convenient to incubate with a very dilute solution of antibody overnight at 4°C instead of for an hour at room temperature). Lectin binding, particularly binding of plant lectins, is generally optimum at around 20°C (i.e. room temperature) and may decrease at lower or higher temperatures. A word of warning, the optimum temperature for lectin binding may sometimes be lower than the usual room temperature and some experimentation is worthwhile. Remarkably, some mammalian lectins appear to bind best at 7°C or 8°C, and minimally at 37°C.

4.1.7 Optimum pH

In our experience, most lectins work well over a wide pH range and using lectin buffer at pH 7.6 should give excellent results with most lectins. However, the pH optimum for lectin binding has not been ascertained in many cases: some lectins are pH dependent, and if problems occur, experimentation is often helpful.

4.1.8 Optimum dilution of lectin

A good, optimal dilution for most lectins in most indirect histochemical methods seems to be 10 μg ml^{-1}. We would recommend that investigators begin experiments using lectin at approximately this concentration, and then adjust conditions to optimize results.

Notable exceptions are PWA, which works well at 1 μg ml^{-1}, and LPA, which may require concentrations as high as 100 μg ml^{-1}.

4.1.9 Storage of lectins and stock solutions

Commercially available lectins are usually supplied as a dry powder, which should be stored either frozen or at 4°C (follow the manufacturer's instructions). In this state most are stable for many years.

When in use, lectins are most conveniently stored as a stock solution of 1 mg ml^{-1} in lectin buffer at 4°C. Again, in this state many are stable for many years. Most native lectins will also keep well for some weeks or months stored in buffer at 'optimal dilution' and at 4°C. The only exceptions to this rule in our experience, are LPA and VVA which are both particularly labile.

4.1.10 Sugar binding characteristics

Although lectins may have the same nominal monosaccharide specificity (i.e. a simple sugar that can inhibit specific red cell agglutination), the precise sugar structure they selectively bind can be enormously different, as described in Section 1.5. Thus, lectins with a nominal specificity for the same monosaccharide (e.g. galNAc) can show a wide range of binding to the same tissue (examples are given in *Figure 3.7*). Do not make the mistake of thinking that, for example, all gal-binding lectins will give the same results – they certainly won't.

4.1.11 Glycotope retrieval: trypsinization and microwaving

The use of trypsinization and microwaving of paraffin sections to reveal carbohydrate structures hidden during fixation and processing is discusssed in Section 3.4.4. Cryostat sections, cell smears and tissue imprints *do not need trypsin or microwave* treatment. Only use trypsinization or microwaving for tissues that have been fixed and processed to paraffin, otherwise the glycoconjugates will wash out and the tissue disintegrate.

4.2 Direct method

The simplest and most straightforward method is the 'direct method'. This makes use of lectin directly conjugated to a fluorescent or enzyme label. A very large range of lectins are commercially available ready conjugated to labels, including the fluorescent tags FITC and TRITC, and enzyme labels including horseradish peroxidase and alkaline phosphatase. It is relatively straightforward to conjugate lectins to enzyme or fluorescent labels. Methods are given in Section 2.4.

The principle of the method is straightforward: the cell/tissue preparation is incubated with the directly labelled lectin. The lectin binds to glycoconjugates expressed by the cells, and binding is visualized by the label directly attached to it. The method is illustrated in *Figure 4.4*.

The direct method has the overwhelming advantage of being very quick and easy to carry out. It may be used in conjunction with any cell or tissue preparation, but is especially recommended for lectin binding to cell suspensions using a fluorescent label such as FITC or TRITC. On tissue sections, it is much less sensitive than the more complex multi-step techniques described later.

In our hands, the use of lectins directly conjugated to the relatively large horseradish peroxidase molecule may sometimes give slightly different staining results to the native lectin, presumably due to the large peroxidase molecule physically interfering with the binding site of the lectin (Brooks *et al.*, 1996). We have experienced no comparable

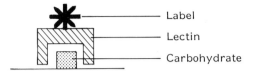

Label
Lectin
Carbohydrate

Figure 4.4:   Direct method: labelled lectin.

problems when using fluorescently labelled lectins or lectins labelled with alkaline phosphatase or biotin. We would therefore recommend their use, or use of a method employing native, unconjugated lectin in preference to peroxidase-labelled lectin.

Methods for detection of fluorescently labelled lectin binding to cell suspensions and alkaline phosphatase-labelled lectin to sections, imprints or smears are given in *Boxes 4.1* and *4.2* as examples. A concentration of 10 µg ml^{-1} lectin, and incubation times of 30 min and 1 h are listed, but the individual worker may need to experiment to obtain optimum results.

Box 4.1: Direct method: fluorescently labelled lectin binding to cell suspensions

1. Prepare a suitable cell suspension in lectin buffer as decribed in *Box 3.1*.
2. Add fluorescently labelled lectin (e.g. FITC or TRITC conjugate) to the cells to give a final concentration of 10 µg ml^{-1}.
3. Incubate the cells with the lectin in an end-over-end mixer for approximately 30 min–1 h at room temperature.
4. Pellet the cells by spinning in a benchtop centrifuge at approximately 300–400 *g* for 5 min.
5. Draw off the supernatant and discard it.
6. Add fresh lectin buffer to wash. Gently resuspend the cells by rocking the tube, or by placing in an end-over-end mixer.
7. Repeat steps 4 and 5 twice more.
8. Place a drop of cells on a clean glass microscope slide using a Pasteur pipette. Cover the drop with a coverslip (see note 1). View under a fluorescence microscope.

Note
1. The coverslip can be sealed around the edges using nail polish. This is particularly useful when oil immersion is used.

Examples of cells in suspension labelled by fluorescently conjugated lectins are given in *Figure 3.5*.

Troubleshooting
1. Lectin binding unsuccessful.The most common reason for lectin not binding to cells in suspension (other than genuine absence of carbohydrates which the lectin might bind to) is the presence of competitively inhibiting glycoconjugates in the solution in which the cells are suspended (e.g. residual glycoconjugates from serum or tissue culture fluid). Ensure that cells have been gently but thoroughly washed in several changes of lectin buffer before incubating with the lectin (follow the method in *Box 3.1*).

(continued)

If enzyme treatment (e.g. trypsin or other proteases) has been used to get cells into suspension, glycoproteins at the cell surface may have been damaged or contaminating glycosidases in the enzyme preparation may have stripped or damaged cell surface glycoconjugates. Allow cells time in culture (12–48 h) to recover from enzyme treatment before lectin binding.

2. Internalization of lectin. Living cells will have a tendency to internalize lectin bound to their surface glycoconjugate. For this reason, there may be an optimum incubation time after which the cells should be examined for lectin binding and photographed, if required. A 30 min incubation time is recommended in the protocol above, but the individual worker may find that a slightly longer or shorter incubation time gives better results.

   If internalization of the lectin is a major problem, cells can be killed prior to lectin binding by fixing briefly in formol saline. Cells are pelleted by spinning at approximately 500 *g* in a bench-top centrifuge for 15 min. They are gently resuspended in formol saline (for recipe see Appendix A) and allowed to incubate in an end-over-end mixer for approximately 30 min. Several washes in lectin buffer are necessary before incubating with the lectin, as any traces of formol saline may interfere with lectin binding.

3. Capping. Lectins will cross-link surface glycans and, as the surface of a living cell is a fluid mosaic, the surface aggregates form patches which may polarize to one end of the cell then pinocytose or shed off. This may be avoided by blocking membrane mobility by briefly incubating cells in the presence of 0.01% (w/v) sodium azide in lectin buffer before lectin binding.

Box 4.2: Direct method: alkaline phosphatase-labelled lectin binding to sections, smears or imprints

1. Prepare sections, imprints or smears as described in Sections 3.4.2, 3.4.3 and 3.4.4.
2. If paraffin sections are to be used, trypsinization or microwaving may be required (see Section 3.4.4).
3. Wash slides in running tap water for approximately 5 min (see Section 4.1.2).
4. Incubate with alkaline phosphatase-conjugated lectin at a concentration of 10 µg ml^{-1} in lectin buffer for 1 h, in a humid chamber, at room temperature.
5. Wash slides 3× in lectin buffer.
6. Incubate with naphthol-AS-BI-phosphate–Fast Red substrate for alkaline phosphatase (see *Box 3.15*) for 10–20 min.
7. Wash in running tap water for approximately 5 min.
8. Counterstain with haematoxylin (*Box 3.20*).
9. Mount in an aqueous mountant (see *Box 3.23*).

Notes

1. Alternative substrates to naphthol-AS-BI-phosphate (see *Boxes 3.14* and *3.16*) may be substituted in step 6. If the product is alcohol soluble, slides must be mounted in aqueous mountant, as listed in step 9. If the product is alcohol insoluble, sections should be dehydrated, cleared (*Box 3.22*) and mounted in a resinous mountant instead. Depending on the colour of the product a counterstain other than haematoxylin, step 8, may be more appropriate (see section 3.5.2).
2. Horseradish peroxidase labelled lectins may be used in place of alkaline phosphatase labelled lectins in step 4. If a peroxidase labelled lectin is used, endogenous peroxidase must first be blocked by incubation with methanol-hydrogen peroxide (Box 3.18). A suitable substrate for peroxidase (described in section 3.5.2) should be substituted for naphthol-AS-BI-phosphate in step 6. If the product is alcohol soluble, slides must be mounted in aqueous mountant, as listed in step 9. If the product is alcohol insoluble, sections should be dehydrated, cleared (Box 3.22) and mounted in a resinous mountant instead.

4.3 Simple indirect method

In this method, the cell or tissue preparation is incubated first with a native, unconjugated lectin, and the lectin binding is detected by a labelled antibody directed against the lectin. The principle of the method is illustrated in *Figure 4.5*.

A range of polyclonal antisera, usually raised in rabbits, against the most commonly used lectins are available commercially, either unlabelled or labelled with fluorescent tags, horseradish peroxidase or alkaline phosphatase (see Sections 2.2 and 2.3). This method is quick and simple to carry out and has the advantage that, as native, unconjugated lectin is used, the potential problem of the label molecule sterically hindering the binding site of the lectin (see Section 4.2) is avoided. This indirect method is more sensitive than the direct method, but less sensitive than the more complex multi-layer methods described later (e.g. in Sections 4.4 and 4.5.3). A slightly more sensitive adaptation of the method is to incubate cell or tissue preparations with unlabelled lectin, then *unlabelled* antibody against it (e.g. unlabelled rabbit anti-lectin) then a labelled second antibody (e.g. labelled swine anti-rabbit).

The method may be used in conjunction with any suitable tissue sections, imprints or smears. Sample protocols for staining cell or tissue preparations for the binding of lectin using the indirect method are given in *Boxes 4.3* and *4.4.*

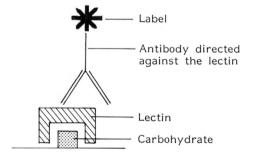

Figure 4.5: Simple indirect method: unconjugated lectin detected by labelled antibody.

Box 4.3: Simple indirect method: sections, smears or imprints stained for the binding of native lectin detected by a fluorescently labelled antibody

1. Prepare sections, imprints or smears as described in Sections 3.4.2, 3.4.3 and 3.4.4.
2. If paraffin sections are to be used, trypsinization or microwaving may be required (see Section 3.4.4).
3. Wash slides in running tap water for approximately 5 min (see Section 4.1.2).
4. Incubate with native, unconjugated lectin at a concentration of 10 µg ml^{-1} in lectin buffer for 1 h in a humid chamber.
5. Wash 3× in lectin buffer.
6. Incubate with fluorescently labelled polyclonal antisera raised in, for example, rabbit against the lectin, diluted 1/100 in lectin buffer, for 1 h.
7. Wash 3× in lectin buffer.
8. Mount in an aqueous mountant (see *Box 3.23*) and view immediately using a fluorescence microscope.

Box 4.4: Simple indirect method: sections, smears or imprints stained for the binding of native lectin detected by a horseradish peroxidase-labelled antibody

1. Prepare sections, imprints or smears as described in Sections 3.4.2, 3.4.3 and 3.4.4.
2. If paraffin sections are to be used, trypsinization or microwaving may be required (see Section 3.4.4).
3. Block endogenous peroxidase by incubating sections in methanol–hydrogen peroxide for 20 min (*Box 3.18*).
4. Wash slides in running tap water for approximately 5 min (see Section 4.1.2).
5. Incubate with native, unconjugated lectin at a concentration of 10 µg ml^{-1} in lectin buffer for 1 h in a humid chamber.
6. Wash 3× in lectin buffer.
7. Incubate with horseradish peroxidase-labelled polyclonal antisera raised in rabbit against the lectin, diluted 1/100 in lectin buffer, for 1 h.
8. Wash 3× in lectin buffer.
9. Incubate with DAB–hydrogen peroxide (see *Box 3.11*) for 10 min, or longer (see note 1).
10. Wash in running tap water for approximately 5 min.
11. Counterstain in haematoxylin (*Box 3.20*).
12. Dehydrate, clear and mount in resinous mountant (see *Boxes 3.22* and *3.23*).

Notes
1. Check the progression of the staining reaction under the microscope and stop it when a good signal to noise ratio is achieved (i.e. a good strong, specific stain against a clean background).
2. Aminoethylcarbazole or chloronaphthol may be used in place of DAB in step 8 (see *Boxes 3.12* and *3.13*). The coloured products of these substrates are alcohol soluble, so slides must be mounted in an aqueous rather than resinous mountant, step 11 (see Section 3.5.2).
3. Alkaline phosphatase-labelled antibody may be used in place of peroxidase-labelled antibody in step 6: here step 2 (blocking endogenous peroxidase) is omitted, and one of the chromogenic substrates given in *Boxes 3.14, 3.15* or *3.16* is used. If the reaction product is alcohol soluble, slides must be mounted in aqueous mountant (see Section 3.5.2).

4.4 APAAP or PAP methods

The alkaline phosphatase–anti-alkaline phosphatase (APAAP) and peroxidase–anti-peroxidase methods are virtually identical (*Box 4.5*). APAAP uses alkaline phosphatase as the label; PAP uses horseradish peroxidase. Both are long, multi-step techniques that, although time consuming because of the large number of steps involved, give good results and are very sensitive. Once very popular, they are now less commonly used and have to a certain extent been superseded by the ABC technique (see Section 4.5.3).

Box 4.5: PAP or APAAP method to demonstrate lectin binding

1. Prepare sections, imprints or smears as described in Sections 3.4.2, 3.4.3 and 3.4.4.
2. If paraffin sections are to be used, trypsinization or microwaving may be required (see Section 3.4.4).
3. For the PAP method only, block endogenous peroxidase by incubating sections in methanol–hydrogen peroxide for 20 min (*Box 3.18*).
4. Wash slides in running tap water for approximately 5 min (see Section 4.1.2).
5. Incubate with native, unconjugated lectin at a concentration of 10 µg ml^{-1} in lectin buffer for 1 h in a humid chamber.
6. Wash 3× in lectin buffer.
7. Incubate with unlabelled rabbit anti-lectin antisera diluted 1/100 in lectin buffer for 1 h.
8. Wash 3× in lectin buffer.
9. Incubate with swine anti-rabbit antibody diluted 1/50 in lectin buffer, 1 h.
10. Wash 3× in lectin buffer.
11. Incubate with PAP or APAAP complex diluted 1/200 in lectin buffer for 1 h.
12. Wash 3× in lectin buffer.
13. For PAP, incubate with DAB–hydrogen peroxide (see *Box 3.11*) for 10 min. For APAAP, incubate with naphthol-AS-BI-phosphate–Fast Red (*Box 3.15*) for 10–30 min (check colour development under the microscope).
14. Wash in running tap water for approximately 5 min.
15. Counterstain with haematoxylin (*Box 3.20*).
16. For PAP, dehydrate, clear and mount in resinous mountant (*Boxes 3.22* and *3.23*). For APAAP, mount in aqueous mountant (*Box 3.23*, see Section 3.5.2).

Notes

1. Aminoethylcarbazole or chloronaphthol may be used in place of DAB in step 13 (see *Boxes 3.12* and *3.13*). The coloured products of these substrates are alcohol soluble, so slides must be mounted in an aqueous rather than resinous mountant, step 12 (see Section 3.5.2).
2. Other chromogenic substrates may be used in place of naphthol-AS-BI-phosphate–Fast Red in step 13 (*Boxes 3.14* and *3.16*).

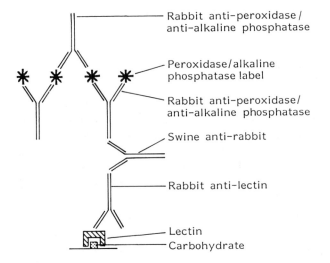

Rabbit anti-peroxidase/
anti-alkaline phosphatase

Peroxidase/alkaline
phosphatase label

Rabbit anti-peroxidase/
anti-alkaline phosphatase

Swine anti-rabbit

Rabbit anti-lectin

Lectin
Carbohydrate

Figure 4.6: Peroxidase–anti-peroxidase (PAP) or alkaline phosphatase–anti-alkaline phosphatase (APAAP) method.

The principle of both techniques is similar, except that one (APAAP) uses alkaline phosphatase as the label and the other (PAP) uses horseradish peroxidase. The cell or tissue preparation is incubated with the unlabelled lectin, then with the unlabelled rabbit antibody against the lectin, then a 'linking' antibody (usually swine anti-rabbit) is added in excess. The next stage is to add the PAP or APAAP complex. This is usually purchased ready made from a commercial supplier, and is a mixture of enzyme label complexed with rabbit antibody directed against the label. The complex traps a relatively high concentration of enzyme label, thus increasing the sensitivity of the method. The complex localizes lectin binding to the cell or tissue preparation through the swine anti-rabbit 'linking antibody' recognizing the rabbit immunoglobulins. The principle of the method is illustrated in *Figure 4.6*.

4.5 Avidin–biotin methods

The affinity of avidin (a protein derived from egg white) for biotin (one of the B group of vitamins) may be exploited in an array of histochemical techniques to detect lectin binding. Streptavidin (derived from *Streptomyces* species) is usually prefered to avidin as it gives a cleaner staining result. Avidin and streptavidin are commercially available conjugated to fluorescent labels, horseradish peroxidase and alkaline phosphatase.

4.5.1 Biotinylated lectin

The simplest avidin–biotin technique relies on the detection of biotinylated lectin by labelled avidin or streptavidin. Many lectins are commercially available ready labelled with biotin. Lectins (even impure, in-house preparations) may be simply and cheaply labelled with biotin in the laboratory: full methods are given in *Box 2.8*. In our hands, biotinylated lectins offer a sensitive, fast and very straight-forward method for detecting carbohydrate residues. The biotin label is a relatively small molecule and does not appear to significantly alter the binding characteristics of the lectin, a problem that may be encountered with peroxidase-conjugated lectins (see Section 4.2). A basic protocol for the detection of carbohydrate residues with biotinylated lectin in cell or tissue preparations is given in *Box 4.6*. The method is illustrated in *Figure 4.7*. The interaction of avidin with biotin may also be exploited in a range of more complex multi-step techniques.

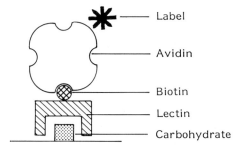

Label

Avidin

Biotin

Lectin

Carbohydrate

Figure 4.7:   Direct avidin–biotin technique: biotinylated lectin.

Box 4.6: Method for the detection of carbohydrate residues using biotinylated lectin and streptavidin peroxidase

1. Prepare sections, imprints or smears as described in Sections 3.4.2, 3.4.3 and 3.4.4.
2. If paraffin sections are to be used, trypsinization or microwaving may be required (see Section 3.4.4.).
3. Block endogenous peroxidase by incubating sections in methanol–hydrogen peroxide for 20 min (*Box 3.18*).
4. Wash slides in running tap water for approximately 5 min (see Section 4.1.2).
5. Incubate with biotin-conjugated lectin at a concentration of 10 μg ml⁻¹ in lectin buffer for 1 h in a humid chamber.

(continued)

6. Wash 3× in lectin buffer.
7. Incubate with streptavidin peroxidase at a concentration of 5 µg ml⁻¹ in lectin buffer for 15–30 min.
8. Wash 3× in lectin buffer.
9. Incubate with DAB–hydrogen peroxide (see *Box 3.11*) for 10 min.
10. Wash in running tap water for approximately 5 min.
11. Counterstain with haematoxylin (see *Box 3.20*).
12. Dehydrate, clear and mount in resinous mountant (see *Boxes 3.22* and *3.23*).

Notes
1. Aminoethylcarbazole or chloronaphthol may be used in place of DAB in step 9 (see *Boxes 3.12* and *3.13*). The coloured products of these substrates are alcohol soluble, so slides must be mounted in an aqueous rather than resinous mountant, step 12 (see Section 3.5.2).
2. Alkaline phosphatase-labelled streptavidin may be used in place of peroxidase-labelled streptavidin: here step 2 (blocking endogenous peroxidase) is omitted, and one of the chromogenic substrates described in *Boxes 3.14, 3.15* or *3.16* is used. If the reaction product is alcohol soluble, slides should be mounted in aqueous mountant (see Section 3.5.2).
3. Streptavidin labelled with a fluorescent label may be used instead of streptavidin peroxidase; here sections are simply mounted using an aqueous mountant (Section 3.5.2) after the washes in step 8.

4.5.2 Biotinylated antibody

Most simply, the binding of unlabelled native lectin can be layered with a biotin-labelled (usually polyclonal rabbit) antibody against it, followed by labelled avidin or streptavidin. The principle of this technique is illustrated in *Figure 4.8* and the method listed in *Box 4.7*. A limited range of biotinylated antibodies are available commercially. A wider range of unlabelled antibodies can be purchased and labelled

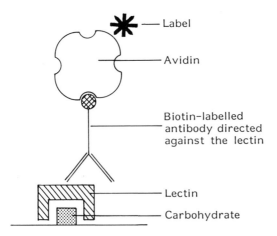

Figure 4.8:   Simple indirect avidin–biotin technique: biotinylated antibody.

with biotin in the laboratory using the simple technique described in *Box 2.8*.

4.5.3 The ABC method

The ABC or avidin–biotin complex technique is a much more complex multi-step technique that has the advantage of being extremely sensitive. Here, unlabelled, native lectin is layered with an unlabelled primary polyclonal rabbit antibody against the lectin. Next, a biotinylated second polyclonal antibody raised in swine against rabbit immunoglobulin is added. The final addition is a labelled pre-formed complex of avidin and biotin mixed in specific proportions so that each avidin molecule has only three of the four possible biotin-binding sites saturated, leaving an extra site free to couple with the biotin label on the secondary antibody. The principle of the technique is illustrated in *Figure 4.9* and the method is listed in *Box 4.8*. The reagents necessary to prepare the ABC complex are commercially available in kit form. The ABC method is exquisitely sensitive and tends to give very good, clean results.

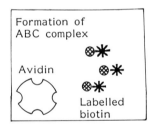

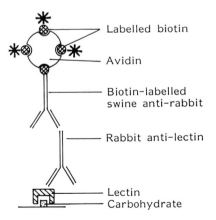

Figure 4.9:   Avidin–biotin complex (ABC) technique.

Box 4.7: Method for detection of native lectin to cell or tissue preparations using biotinylated antibody

1. Prepare sections, imprints or smears as described in Sections 3.3.2, 3.3.3 and 3.3.4.
2. If paraffin sections are to be used, trypsinization or microwaving may be required (see Section 3.3.4).
3. Block endogenous peroxidase by incubating sections in methanol–hydrogen peroxide for 20 min (*Box 3.18*).
4. Wash slides in running tap water for approximately 5 min (see Section 4.1.2).
5. Incubate with native, unconjugated lectin at a concentration of 10 µg ml^{-1} in lectin buffer for 1 h in a humid chamber.
6. Wash 3× in lectin buffer.
7. Incubate with biotinylated rabbit polyclonal antibody directed against the lectin, diluted 1/100 in lectin buffer for 1 h.
8. Wash 3× in lectin buffer.
9. Incubate with streptavidin peroxidase at a concentration of 5 µg ml^{-1} in lectin buffer for 30 min.
10. Wash 3× in lectin buffer.
11. Incubate with DAB–hydrogen peroxide (see *Box 3.11*) for 10 min.
12. Wash in running tap water for approximately 5 min.
13. Counterstain with haematoxylin (see *Box 3.20*).
14. Dehydrate, clear and mount in resinous mountant (see *Boxes 3.22* and *3.23*).

Notes
1. Aminoethylcarbazole or chloronaphthol may be used in place of DAB in step 11 (see *Boxes 3.12* and *3.13*). The coloured products of these substrates are alcohol soluble, so slides must be mounted in an aqueous rather than resinous mountant, step 14 (see Section 3.5.2).
2. Alkaline phosphatase-labelled streptavidin may be used in place of peroxidase-labelled streptavidin (step 9): here step 2 (blocking endogenous peroxidase) is omitted, and one of the chromogenic substrates given in *Boxes 3.14, 3.15* or *3.16* is used. If the reaction product is alcohol soluble, slides must be mounted in aqueous mountant (see Section 3.5.2).
3. Streptavidin labelled with a fluorescent label may be used instead of streptavidin peroxidase (step 9); here sections are simply mounted using an aqueous mountant (Section 3.5.2) after the washes in step 10.

Box 4.8: Method for detection of lectin binding to cell or tissue preparations using the ABC method

1. Prepare sections, imprints or smears as described in Sections 3.4.2, 3.4.3 and 3.4.4.
2. If paraffin sections are to be used, trypsinization or microwaving may be required (see Section 3.4.4).
3. Block endogenous peroxidase by incubating sections in methanol–hydrogen peroxide for 20 min (*Box 3.18*).
4. Wash slides in running tap water for approximately 5 min (see Section 4.1.2).
5. Incubate with native, unconjugated lectin at a concentration of 10 µg ml^{-1} in lectin buffer for 1 h in a humid chamber.
6. Wash 3× in lectin buffer.
7. Incubate with unlabelled rabbit anti-lectin antisera diluted 1/100 in lectin buffer for 1 h.
8. Wash 3× in lectin buffer.

(*continued*)

9. Incubate with biotinylated swine anti-rabbit diluted 1/400 in lectin buffer, 1 h.
10. Wash 3× in lectin buffer.
11. Incubate with peroxidase–ABC complex, prepared according to the manufacturer's instructions for 30 min.
12. Wash 3× in lectin buffer.
13. Incubate with DAB–hydrogen peroxide (see *Box 3.11*) for 10 min.
14. Wash in running tap water for approximately 5 min.
15. Counterstain with haematoxylin (see *Box 3.20*).
16. Dehydrate, clear and mount in resinous mountant (see *Boxes 3.22* and *3.23*).

Notes

1. Aminoethylcarbazole or chloronaphthol may be used in place of DAB in step 13 (see *Boxes 3.12* and *3.13*). The coloured products of these substrates are alcohol soluble, so slides must be mounted in an aqueous rather than resinous mountant, step 12 (see Section 3.5.2).
2. Alkaline phosphatase-labelled ABC may be used in place of peroxidase-labelled ABC in step 11: here step 2 (blocking endogenous peroxidase) is omitted, and one of the chromogenic substrates given in *Boxes 3.14*, *3.15* or *3.16* is used. If the reaction product is alcohol soluble, slides must be mounted in aqueous mountant (see Section 3.5.2).

4.6 A word about kits

Some commercial suppliers sell reagents for the more complex methods (e.g. ABC) in kit form. In our experience, these are generally a very good idea. Reagents are usually supplied in dropper bottles and instructions are very clear and simple (e.g. 'mix one drop of solution A in 5 ml buffer and leave for 10 minutes'). Results are good. Such kits are 'idiot proof' and especially recommended for the novice. They tend to work out slightly more expensive than if reagents are purchased separately.

4.7 Controls

As with all immunocytochemistry, it is vital to include appropriate positive and negative controls in every batch of staining.

4.7.1 Positive controls

The most suitable positive control is a cell or tissue preparation that is known to be positive for the binding of the particular lectin. Always include a positive control with every batch of staining; this is especially important when one is staining consecutive batches of slides at different times (e.g. on consecutive days) to assess the comparability of staining intensity from batch to batch.

4.7.2 Negative controls

The simplest negative control is simply to 'stain' one or more cell or tissue preparations, omitting the lectin from the staining protocol. The cell or tissue preparation should show no staining in the absence of the lectin.

4.7.3 Confirmation of binding specificity

It is essential when performing cytochemistry for lectin binding to confirm the specificity of binding through competitive inhibition by the appropriate simple sugar. There are several ways of doing this: the simplest is to incubate the lectin with the cell or tissue preparation in the presence of a 0.1 M solution of the appropriate monosaccharide (for example, if staining for the binding of Con A, make the lectin solution up in a 0.1 M man or, even better, 0.1 M α-D-methyl-mannopyranoside solution in lectin buffer). Alternatively, following incubation of the cell or tissue preparation with the lectin, attempt to displace it by washing in a 0.1 M solution of the monosaccharide in lectin buffer for 10–30 min.

Note. As lectins prefer to bind to complex sugars rather than monosaccharides, you might expect a *diminution* of staining rather than its complete abolition using these approaches. In our hands, however, competitive inhibition using monosaccharide is often complete.

Troubleshooting.
1. Weak staining/no staining. Poor lectin binding can be due to a number of factors:

   - When using paraffin sections, the most likely cause is sequestration/loss of glycoconjugates during fixation and processing. Glycolipids will be lost during processing (see Section 3.4.4.) and this is irreversible. Glycoproteins may be damaged or hidden and this can usually be reversed by trypsinization or microwaving. The methods are given in *Boxes 3.9* and *3.10*.
   - When using frozen sections, imprints or smears, rapid fixation in acetone can elute glycolipids. If this seems to be the problem, omitting the fixation step altogether, or rapidly fixing in

formalin instead of acetone may help. We have sometimes found when using 'fresh' cell or tissue preparations, that soluble glycoconjugates may elute into solution during incubation with the lectin and competitively inhibit lectin binding. We have overcome this by rinsing slides in running tap water or buffer for 5–15 min prior to staining.

- The most common cause of weak lectin staining is simply inappropriate staining conditions. We would recommend carrying out a 'chequerboard of dilutions' for all reagents prior to use, to determine the optimum conditions. Generally, a higher concentration of lectin will increase sensitivity, although background staining may become a problem. The more complex, and sensitive, multi-step techniques, such as the ABC method, will sometimes give good results when simpler, and less sensitive, techniques give disappointing results. Another way of increasing sensitivity is to increase incubation time with the lectin (e.g. overnight instead of 1 h) or to increase the temperature (e.g. 37°C instead of room temperature). This is true for most lectins, though some do have unpredictable temperature optima (e.g. lectins have been described whose binding is optimal at 7 or 8°C and minimal at 37°C). It is often worth experimenting.

2. High background staining. The ideal is to get good, strong specific staining against a completely clean, negative background. This doesn't always happen! In traditional immunocytochemistry, incubation of cell or tissue preparations in the presence of 1–5% normal serum (e.g. normal swine serum) will clean up dirty staining very effectively. This will not work with lectin histochemistry as the copious quantities of glycoconjugates present in serum will competitively inhibit specific staining. In lectin histochemistry, incubate cell or tissue preparations in the presence of 1–5% (w/v) BSA (or another non-glycosylated protein). We have found that this is effective and, indeed, we routinely make up lectin solutions for use in 1% (w/v) BSA in lectin buffer.

3. Sections float away during staining. This is a common problem. The usual cause is poor quality sections; any nicks, tears or creases will act as a focus from which the section will slowly float away from the glass slide. Another cause is dirty, or greasy, microscope slides; the section will not adhere well to a dirty slide. Slides should be taken fresh from the packet when required, should be handled carefully by the edges, and fingerprints on the glass should be avoided. During long complex multi-step methods even the finest quality sections can sometimes become dislodged from the glass slide due to the repeated soaking and vigorous washing. Pre-coating slides with an adhesive such as poly-L-lysine (see *Box 3.3*) is highly recommended.

4. When using resinous mountant, preparations look cloudy, or bubbles form on storage. Both of these problems are due to water contamination of the 99% alcohol or xylene used in dehydrating and clearing sections. Sections should be soaked overnight in xylene to loosen the coverslips and wash away the mountant. They should then be rehydrated through *fresh* graded alcohols (e.g. 99%, 95%, 70%) then carefully dehydrated, cleared in *fresh* xylene and mounted.

4.8 Some ideas on choice of lectins

It is best to commence using a cheap, well documented lectin that binds to a wide variety of tissues. Con A is a splendid starter, as it is cheap and binds to most cells and tissues. UEA-I is also very easy to use and gives good results on human blood vessel endothelium (try sections of striated muscle; see *Figure 3.7*). PNA binds to a variety of tissues – it is especially good on human kidney (see *Figure 3.5*). Some simple, and successful, methods are given in Section 4.11.

4.9 Hazardous lectins

RCA, APA and VAA are extremely toxic and should be avoided if possible. Other, safer lectins with similar carbohydrate specificity are available. Comprehensive hazard data for most lectins are not available, and so they should be treated with reasonable care in the laboratory: dust and aerosols should obviously be avoided.

4.10 Which tissues to use

To get started, you need a readily available, reproducible source which binds a wide variety of lectins. Kidney is very good for this, even fresh pig kidney from a supermarket. Epithelial rich tissue is generally successful. Pituitary (for glycosylated hormone precursors) and salivary glands (different mucins) give an interesting pattern, particularly with lectins which bind to O-linked carbohydrate residues.

4.11   Foolproof methods for the beginner

4.11.1 UEA-I binding to human endothelium

U. europaeus (gorse) has two main isolectins: UEA-I binds alpha-L-fuc and UEA-II binds glcNAc.

UEA-I is a potentially useful tool for identification of human endothelial cells in tissue sections and in culture (it generally does not work on animal endothelia). It also stains tumours of endothelial origin intensely. UEA-I can give very strong staining, particularly on fresh tissues or after trypsin digestion in paraffin sections. A simple method is given in *Box 4.9*. UEA-I binding to human endothelium using this method is illustrated in *Figure 3.7*.

Box 4.9: Direct peroxidase method for the binding of UEA-I to paraffin sections

1. Prepare paraffin sections or frozen (cryostat) sections of human striated muscle, or any other endothelium-rich tissue as described in Sections 3.4.3 or 3.4.4.
2. Block endogenous peroxidase by soaking in methanol–hydrogen peroxide (see *Box 3.18*) for 20 min.
3. Wash in running tap water for 5 min.
4. If paraffin sections are used, they should be trypsinised (see *Box 3.9*) 20 min.
5. Wash in running water, 5 minutes.
6. Incubate with peroxidase-conjugated UEA-I, at a concentration of 10μg/ml in lectin buffer, for 1 hour, in a humid chamber.
7. Wash 3× in lectin buffer.
8. Incubate with DAB-hydrogen peroxide, (see Box 3.11) 10 minutes.
9. Wash in running tap water, 5 minutes.
10. Counterstain in haematoxylin (see Box 3.20).
11. Dehydrate, clear and mount in resinous mountant (see Boxes 3.22 and 3.23).

Using this method the endothelium of most capillaries, arterioles and venules will be intensely stained.

4.11.2 PNA binding to tubules of the kidney

Kidney (either human kidney if you have access to clinical material, or animal kidney from a supermarket or butcher's shop) is an excellent tissue on which to screen lectins as there appears to be very variable expression of carbohydrate structures at different points along the tubules. The result is that most lectins will give some staining on kidney tissue. PNA is a good lectin to start with: it binds intensely to the luminal surface of some tubules in frozen sections or in paraffin sections following trypsin digestion (see *Figure 3.5*). In our hands, the addition of 1% (w/v) BSA to all staining solutions significantly improves the cleanness of the final staining product.

Other lectins (e.g. DBA, WGA, SBA, and HPA) can all be substituted in the method in *Box 4.10*, and will give different (but just as impressive) staining results.

Box 4.10: Direct alkaline phosphatase method for PNA binding to paraffin sections of kidney

1. Prepare paraffin sections or frozen (cryostat) sections of human kidney as described in Sections 3.4.3 or 3.4.4.
2. If paraffin sections are used, they should be trypsinized (see *Box 3.9*) for 20 min.
3. Wash in running tap water for 5 min.
4. Incubate with alkaline phosphatase-conjugated peanut lectin at a concentration of 10 µg ml^{-1} in 1% (w/v) BSA in lectin buffer, for 1 h.
5. Wash 3× in lectin buffer.
6. Incubate with naphthol phosphate–New Fuchsin (see *Box 3.16*) for 10–30 min.
7. Wash in running tap water for 5 min.
8. Counterstain in haematoxylin (see *Box 3.20*).
9. Dehydrate, clear and mount in resinous mountant (*Boxes 3.22* and *3.23*).

4.12   The way forward: characterization of complex carbohydrates

Lectin histochemistry is really a first step. Binding of lectins to tissue sections can demonstrate differences or changes in glycosylation, but the characterization of the carbohydrates involved is more difficult.

Competitive inhibition of lectin binding by mono- and oligosaccharides will give a superficial idea of the sugar structures recognized (see Section 4.7.3). Further information can be derived from experiments with sequential enzyme digestion. If, for example, a tissue section stains strongly for the binding of LFA, one can conclude that carbohydrate structures rich in terminal sialic acid are being expressed. This can be confirmed by digestion of an adjacent section with the enzyme neuraminidase (which cleaves terminal sialic acid) followed by LFA binding. Now a complete absence of staining is observed (this is illustrated in *Figure 3.7*) confirming that the terminal monosaccharide was sialic acid. The immediate sub-terminal monosaccharide can then be identified by staining for the binding of a panel of lectins with different monosaccharide specificity. If, for example, the sub-terminal monosaccharide is identified as man (e.g. by positive Con A binding), the man can be digested by a mannosidase and the process may be repeated to identify the next monosaccharide in the chain and so on. A typical method for exoglycosidase digestion is given in *Box 4.11*.

Box 4.11: Neuraminidase digestion of terminal sialic acid

Terminal sialic acid may be detected by the binding of a number of lectins, including LPA, SNA, MAA and LFA. If a cell or tissue preparation is positive for the binding of one of these lectins, this indicates the presence of terminal sialic acid. Sialic acid may be cleaved by pre-treatment with neuraminidase as follows.

1. Prepare cell or tissue samples as described in Sections 3.4.2, 3.4.3 or 3.4.4.
2. Incubate with 0.1 units/ml neuraminidase in lectin buffer, at 37°C, for 30 min–1 h (note 1).
3. Wash 3× in lectin buffer.

Subsequent LFA, LPA or SNA binding should now be abolished.

Notes
1. Different neuraminidase enzymes have different pH optima. Refer to the manufacturer's instructions and adjust the pH of the lectin buffer accordingly.
2. A word of warning: Plendl *et al.* (1989) have shown abolition of staining using simply buffer in the absence of neuraminidase enzyme, which casts doubt on the validity of results using this approach.
3. LFA, MAA, LPA and SNA can give good results, but, unlike most lectins, LPA is extremely labile: once dissolved in buffer to form a stock solution it will be stable at 4°C for only a few weeks (most lectins are highly stable and can be stored for years with no loss of activity).
4. LPA also yields optimum results when used at an unusually high concentration. Most lectins will work well at approximately 10 μg ml^{-1}; LPA should be used at approximately 100 μg ml^{-1}.
5. In theory, the sequential enzyme digestion approach should work very well, but in practice, as described below, owing to the complexity and heterogeneity of carbohydrate structures expressed by cells and tissues, results are often ambiguous and difficult to interpret.
6. To gain meaningful results it is essential to use the purest quality (and therefore usually the most expensive) exoglycosidase enzyme preparations, to limit the effects of any contaminating enzyme.

Sequential enzyme digestion for analysis of tissue glycoconjugates is, however, much more difficult than it might appear. Difficulties are encountered because of the huge range of structures simultaneously present in tissues: unless one structure predominates, a change in intensity of lectin binding before vs. after digestion by an enzyme may not be obvious.

This type of investigation can be interesting, but meaningful analysis really needs extraction, isolation and then biochemical or physical analysis of the glycans. This may involve expensive and time-consuming procedures such as mass spectrometry, nuclear magnetic resonance and computer-assisted comparison of the results of these studies with previously characterized glycans.

References

Brooks SA, Lymboura M, Schumacher U, Leathem AJC. (1996) Histochemistry to detect *Helix pomatia* lectin binding – methodology makes a difference. *J. Histochem. Cytochem.* **44**, 519–524.

Plendl J, Schönleber B, Schmahl W, Schumacher U. (1989) Comparison of the unmasking of lectin receptors by neuraminidase and by enzyme free buffer alone. *J. Histochem. Cytochem.* **37**, 1743–1744.

5 Electron Microscopic Methods for Demonstration of Lectin Binding Sites

5.1 Introduction

Lectin histochemistry at the electron microscope level can reveal the precise distribution of carbohydrates in the sub-cellular compartment, for example within the cisternae of the Golgi apparatus. The principles of the techniques are similar to those used at light microscope level, but depend on the presence of an electron dense (rather than fluorescent or coloured) label. As in light microscopy, both direct and indirect methods are commonly used, although there are more limitations at the ultrastructural level due to the nature of tissue processing for electron microscopy. Fixation and processing are more critical for electron microscopy, and more time and effort is generally required. Before starting any electron microscopical work, the question should be asked whether localization to the sub-cellular level is really required or whether other techniques (e.g. thin sectioning of plastic-embedded specimens or confocal microscopy) might not suffice for localization purposes.

5.2 Choice of label

5.2.1 Gold and horseradish peroxidase

Electron microscope histochemistry has fundamentally changed in recent years. Whereas previously the electron dense marker of choice was often DAB (after reaction with the enzyme label horseradish peroxidase) or ferritin (see *Figure 5.1* for an example of ferritin labelling), colloidal gold particles are now commonly used. Colloidal gold gives a much more specific localization of labelling in comparison to DAB. Gold particles of 5–20 nm in diameter are commonly used.

5.2.2 Double labelling

The availability of different sizes of gold particle allows the co-localization of two different carbohydrate structures within the same sample by choosing a large diameter gold label for one lectin and a significantly smaller diameter gold label for the other. However, potentially there may be problems. The smaller the gold particle size, the more favourable the label:lectin size ratio. Very large diameter gold particles tend to give less specific localization since the lectin forms a very minor part of the lectin/gold complex. We have carried out experiments using various labels for lectins and measuring respiratory bursts of human monocytes: the binding of gold-labelled lectins resulted in almost no cell response. On the basis of these experiments, we feel that the results of lectin binding experiments on living cells using gold-labelled lectins should be interpreted with some care.

5.3 Direct and indirect detection methods

As with lectin binding for light microscopy, two choices exist when using gold or peroxidase as a label for visualization of lectin binding: either a direct or an indirect method. In the direct method, lectins are conjugated directly to the electron dense label. In the indirect method,

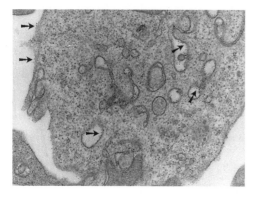

Figure 5.1: Binding and uptake of ferritin-labelled WGA into human monocytes. The living cells were incubated with the lectin to observe the internalization. Note the accumulation of ferritin granules within the lumen of the vesicles. Magnification 5000x.

biotinylated lectin is visualized by the binding of a second layer (avidin or streptavidin) labelled with gold or peroxidase. The basic principles of both methods are exactly the same as for light microscopy (Sections 4.2 and 4.5.1). A third choice is to use lectin labelled with digoxigenin, then detect its binding through use of an anti-digoxigenin antibody and then a second, peroxidase or gold-labelled antibody. The principle of this method is illustrated in *Figure 5.2*.

We would recommend the use of biotinylated lectin in a simple indirect method. A wide range of lectins are available commercially ready-labelled with biotin, and biotinylation of unconjugated lectin is a simple procedure (*Box 2.8*) to carry out in the laboratory. The method is then extremely flexible: purchase of a small stock of avidin or streptavidin labelled with gold or peroxidase is all that is required to investigate the binding of a virtually limitless range of lectins. A more limited range of lectins is commercially available labelled with gold or digoxigenin. Labelling lectins 'in house' with colloidal gold is quite complex and involves isoelectric focusing of the lectin to determine pI and ultracentrifugation of the colloidal gold. Antibodies against digoxygenin, and avidin- or streptavidin-labelled gold are readily available, for example from Sigma Chemical Co. and Boehringer Mannheim.

5.4 Pre-embedding and post-embedding techniques

In addition to direct and indirect detection methods, pre- and post-embedding techniques are also distinguished. There are few reports

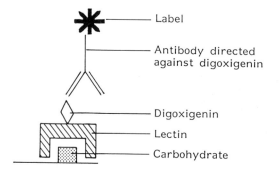

Figure 5.2:   Use of digoxigenin-labelled lectin to detect carbohydrate residues.

in the recent literature of the use of pre-embedding methods for lectin histochemistry, where, as the name implies, the tissues are exposed to the lectin prior to embedding and sectioning. The main reason for this is that pre-embedding techniques have lost popularity owing to the sometimes patchy staining resulting from uneven binding of the lectin because of problems of lectin penetration. In the now more commonly used post-embedding methods, tissues are first processed and sectioned, and then the lectins are applied to the ultra-thin sections. We would generally recommend using post-embedding techniques as, in our hands, they give superior results.

5.4.1 Pre-embedding techniques

Pre-embedding methods are restricted in their suitability because of the difficulties of penetration of lectins and lectin–marker complexes into cells. This is because the molecular size of lectins is relatively large, especially when conjugated to a marker such as horseradish peroxidase, or even colloidal gold particles. For this reason, a permeabilization step may be incorporated. The pre-embedding methods may be utilized in either the direct or indirect methods outlined in Section 5.3.

5.4.2 Post-embedding techniques

Post-embedding methods are the method of choice for electron microscope demonstration of lectin binding sites. One potential problem is that the tissue processing necessary for electron microscopy may in some cases destroy carbohydrate integrity and thus reactivity for lectins. It is therefore a good idea to check whether the lectin ligand in question requires a particular type of fixation by trying a range of different fixatives to determine which gives the optimum lectin reactivity. Carbohydrates are, however, relatively robust, and tend to be less damaged by aggressive glutaraldehyde fixation than protein epitopes.

Use of hydrophilic resins such as Lowicryl-K4M obviates the need for etching in post-embedding methods, though use of Araldite results in better structural preservation. Post-embedding techniques may be employed in conjunction with either the direct or indirect methods outlined in Section 5.3. The post-embedding methods are popular because a range of lectins can be applied to serial sections, and this is not possible with pre-embedding methods.

5.5 Methods for demonstration of lectin binding at the EM level

Methods are given in *Boxes 5.1–5.4* for demonstrating gold-labelled lectin binding to cultured cells at the EM level using both pre-embedding and post-embedding techniques. The methodology can be readily adapted for use with other lectins, or with other specimens, and the investigator is encouraged to experiment.

Box 5.1: Preparation of fixative for cell samples

Fixation of cultured cells for localization of lectin binding at the EM level is carried out in 4% (w/v) paraformaldehyde and 0.5% (v/v) glutaraldehyde in 0.1 M PBS, pH 7.4.

Making up the paraformaldehyde solution is a critical step:
1. Dissolve the paraformaldehyde by heating it in a small volume of PBS at about 65–70°C using a heated magnetic stirrer.
2. Add 1 M sodium hydroxide drop by drop until the pH reaches 7.4.
3. Add glutaraldehyde to give a concentration of 0.5% (v/v).
4. Add the rest of the PBS so that the final concentration of paraformaldehyde is 4% (w/v).

Fixatives should preferably be used ice-cold. Although aldehyde fixation is preferred, numerous permutations of the combination of paraformaldehyde and glutaraldehyde have been used (Mitchell and Schumacher, 1996), the optimal permutation for any particular cell sample and lectin binding method must be determined by experimentation.

Box 5.2: Cell maintenance prior to labelling

1. Cell lines should be maintained in media as recommended by the supplier in a humidified atmosphere of 5% carbon dioxide and 95% air at 37°C.
2. Media should be changed approximately every 3 days, until the cultures are confluent. They should then be harvested.
3. To harvest, cells should be gently scraped from the culture flask using a rubber scraper.
   NB. Do not use digestive enzymes such as trypsin to detach cells from the plastic as this will disrupt their cell surface glycosylation and drastically alter lectin binding results.
4. Cells pelleted by gentle centrifugation at 500 *g* in a bench-top centrifuge for 10–15 min.

Box 5.3: Pre-embedding technique for lectin binding

1. Fix cell pellets by immersion in 4% (w/v) paraformaldehyde and 0.5% (v/v) glutaraldehyde in 0.1 M PBS, pH 7.4 (see *Box 5.1*) for 15 min.
2. After a brief wash in PBS, react cell samples with 0.05% (w/v) sodium borohydride in PBS to reduce background reactivity.
3. Wash pellets in PBS then treat with 0.1% (w/v) saponin in PBS for 30 min to permeabilize the cell membranes prior to incubation with the lectin.
4. Incubate the cell samples with gold-labelled lectin (10 nm diameter gold particles) at a concentration of 10 µg ml^{-1} in lectin buffer for 4 h at room temperature on a rocking table.
   NB. The conditions given above should work well for most lectins, however, the investigator is encouraged to determine optimum conditions for any particular lectin and cell preparation using a 'chessboard titration'. For example, three dilutions of the lectin (5 µg ml^{-1}; 10 µg ml^{-1}; 20 µg ml^{-1}) may be incubated at each of three times (2 h, 4 h and 6 h).
5. Wash cell pellets briefly in PBS.
6. Post-fix the cells in 1% (w/v) glutaraldehyde in PBS for 15 min, followed by washes in PBS.
7. Fix the cell pellets further in 0.5% (w/v) osmium tetroxide in PBS and 0.5% (w/v) potassium ferricyanide in PBS before block staining in 1% (w/v) uranyl acetate in PBS for 30 min.
8. After a brief wash in distilled water, embed the cell pellets in 1% (w/v) agar (see note 1). After cooling to a temperature just above the setting point of the agar, embed the cell pellets in the agar, allow it to set, and cut the small part of the agar containing the cells into small pieces.
9. Dehydrate the pellets in 50% (v/v) acetone in distilled water for 15 minutes, then 75% (v/v) acetone in distilled water for 15 min, 95% (v/v) acetone in distilled water for 20 min, 100% ethanol for 15 min, and finally 100% acetone (three changes of 20 min duration each).
10. Infiltrate the pellets in a 1:1 mixture of acetone and Araldite for 18 h, and then in neat Araldite firstly for 4 h and then for 18 h before polymerizing in Araldite in gelatin capsules for 18 h at 60°C.

Note

1. Difco Agar Noble should be used as it is the most electron lucent type. It should be prepared by dissolving the agar at a concentration of 1 g/100 ml PBS and heating in a boiling water bath until the agar dissolves.

   Use of this method enables visualization of lectin binding sites on the cell surface as illustrated in *Figure 5.3*.

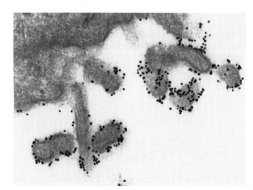

Figure 5.3: Binding of gold labelled HPA to human breast cancer cells (MCF7 cell line).

Pre-embedding procedure. Note the intensive labelling of microvilli at the apex of the cells. Magnification 6500x.

Box 5.4: Post-embedding technique for lectin binding

Embedding using LR White resin.
1. Fix cell pellets in 1% (v/v) glutaraldehyde in PBS for 15 min, followed by washes in PBS.
2. Transfer the cell pellets to 0.5% (w/v) osmium tetroxide in PBS and 0.5% (w/v) potassium ferricyanide in PBS (see note 2).
3. Embed in 1% (w/v) agar (see note 1). After cooling to a temperature just above the setting point of the agar, embed the cell pellets in the agar, allow to set, and cut into small pieces.
4. Dehydrate the pellets in 50% (v/v) ethanol in distilled water for 15 min, 75% (v/v) ethanol in distilled water for 15 min, 95% (v/v) ethanol in distilled water for 20 min and finally three changes of 20 min duration in 100% ethanol.
5. Infiltrate the pellets with a 1:1 mixture of 100% acetone and LR White resin for two 1 h changes, then a 2:1 mixture of LR White resin and acetone for two changes of 1 h duration.
6. Transfer to LR White monomer for two changes of firstly 1 h duration, and then of 18 h, followed by two further changes of 1 h duration.
7. Polymerize in gelatin capsules in LR White resin for 18 h at 60°C.
8. For LR Gold, dehydrate the pellets in 50% (v/v) methanol in distilled water for 15 min, then 75% (v/v) methanol in distilled water for 15 min at −20°C, then in 90% (v/v) methanol in distilled water at −20°C and finally three changes of 20 min duration in 100% methanol at −20°C.

Embedding using LR Gold resin. Follow steps 1–4 as above, then:
5. Expose the pellets to the LR Gold monomer for 2 h, then monomer for 18 h, then a 1:1 mixture of monomer and 0.5% benzoin methyl ether for 4 h, then for 18 h, all at −20°C.
6. Carry out polymerization in gelatin capsules with a 1:1 mixture of fresh LR Gold monomer and 0.5% benzoin methyl ether. To do this, place the gelatin capsules in the wells of a clear, plastic 96-multiwell assay plate on a stand 9 cm above a 20 cm fluorescent light strip (6 W) for 24 h at −20°C.

Sectioning.
1. Cut ultrathin sections and mount them on nickel grids. Treated with blocking buffer containing 99 ml PBS, 1 mg ml^{-1} BSA, 0.1% (w/v) gelatin and 50 mM glycine.
2. Incubate sections on a rocking table with lectin–gold conjugate (10 nm diameter gold particles) at a concentration of 10 µg ml^{-1} in blocking buffer (for recipe see Appendix A), and as determined by previous experiments for 4 h.

Notes
1. Difco Agar Noble should be used as it is the most electron lucent type. It should be prepared by dissolving the agar at a concentration of 1 g 100 ml^{-1} PBS and heating in a boiling water bath until the agar dissolves.
2. Osmification should be omitted if membrane-associated glycosylation is to be studied.

5.6 Controls

5.6.1 Negative controls

In both pre- and post-embedding methods, as a negative control, the lectin stage should be omitted.

5.6.2 Confirming sugar specificity

The lectin is pre-incubated with the appropriate monosaccharide. An equal volume of lectin at a concentration of 20 μg ml^{-1} is mixed with an equal volume of 0.6 M sugar solution to give a final concentration of 10 μg ml^{-1} lectin and 0.3 M sugar. This mixture is incorporated in place of the standard lectin solution in the methods described in *Boxes 5.3* and *5.4*.

Notes.

1. Higher concentrations of aldehyde fixation are permitted with lectin binding methods compared to antibody localization work. However, there is inevitably a compromise between achieving structural integrity and lectin binding intensity.
2. Successful use of the low temperature resin Lowicryl-K4M has been reported (Aviles *et al.*, 1992; Kasper and Migheli, 1993).
3. Ultrathin sections are normally mounted on Formvar-coated nickel grids, though in our experience it is not necessary to coat grids.
4. In pre-embedding methods relatively thick sections can be cut on a Vibratome and reacted with lectins after which they can be embedded. This has an advantage in that lectin binding can occur before any possible deleterious effects of dehydration, or embedding have taken place.

Troubleshooting.

1. Poor penetration of lectin when using pre-embedding techniques. Such problems may be largely overcome by use of detergents like saponin (Yamawaki *et al.*, 1993) or proteolytic enzymes to permeabilize cell membranes. Such procedures will, of course, damage the proteins which carry the carbohydrate residues and may be inappropriate depending on the expected location of the lectin ligand. They should be used with care. Use of a post-embedding technique may be more appropriate.
2. High background labelling. Fixation with aldehydes may produce high background labelling and some authorities claim that treatment of sections with 0.05% sodium borohydride decreases background labelling in antibody localization methods (Weber *et al.*, 1978; Willingham, 1983), and, in our experience, increases lectin binding (Mitchell *et al.*, 1995). It has been suggested that this treatment reduces the number of Schiff bases which arise during the fixation process.

Some workers include treatment with 0.5 M ammonium chloride in PBS for 1 h in post-embedding methods to amidimate free aldehyde groups to reduce background reaction (Castells *et al.*, 1992).

References

Avilés M, Martinéz-Menárguez JA, Castells MT, Madrid JF, Ballesta J. (1992) Cytochemical characterisation of oligosaccharide side chains of the glycoproteins of rat zona pellucida: an ultrastructural study. *Anat. Rec.* **239**, 137–149.

Castells MT, Balesta J, Madrid JF, Martinéz-Menárguez JA, Avilés M. (1992) Ultrastructural localization of glycoconjugates in human bronchial glands: the subcellular organisation of N- and O-linked oligosaccharide chains. *J. Histochem. Cytochem.* **40**, 265–274.

Kasper M, Migheli A. (1993) LR Gold and LR White embedding of lung tissue for immunoelectron microscopy. *Acta Histochem.* **95**, 221–227.

Mitchell BS, Schumacher U. (1996) Electron microscopic methods for demonstration of lectin binding sites. In: *Lectins in Medical Research* (eds JM Rhodes, JD Milton). Humana Press, Totowa NJ, in press.

Mitchell BS, Vernon K, Schumacher U. (1995) Ultrastructural localization of *Helix pomatia* agglutinin (HPA)-binding sites in human breast cancer cell lines and characterization of HPA-binding glycoproteins by Western blotting. *Ultrastruct. Pathol.* **19**, 51–59.

Weber K, Rathke PC, Osborn M. (1978) Cytoplasmic microtubular images in glutaraldehyde-fixed tissue culture cells by electron microscopy. *Proc. Natl Acad. Sci. USA* **75**, 1820–1824.

Willingham MC. (1983) An alternative fixation-processing method for cytoplasmic antigens. *J. Histochem. Cytochem.* **31**, 791–798.

Yamawaki M, Zurbriggen A, Richard A, Vandevelde M. (1993) Saponin treatment for *in situ* hybridisation maintains good morphological preservation. *J. Histochem. Cytochem.* **41**, 105–109.

6 Analysis of Glycoproteins by Electrophoretic Techniques with Western Blotting using Lectins

6.1 Limitations of electrophoretic techniques

6.1.1 Glycoproteins, glycolipids and glycosaminoglycans

The principal glycoconjugates recognized by lectins in cell or tissue samples are glycolipids, glycoproteins, glycosaminoglycans (GAGs) and polysaccharides such as glycogen. Of these, glycoproteins are most easily analysed by the techniques of SDS–PAGE and IEF with Western blotting and subsequent carbohydrate detection by lectins.

6.1.2 Types of glycoproteins

Carbohydrate residues of glycoproteins are the major binding partners for lectins in cell and tissue preparations. They can be either membrane glycoproteins, secretory products (glycoproteins of the serum including immunoglobulins; mucins from, for example, goblet cells) or parts of the extracellular matrix and connective tissues.

Membrane and serum glycoproteins are generally N-linked glycans and fall in the molecular weight range of approximately 20–120 kDa. They are readily analysed by SDS–PAGE and IEF. Mucus type glycoproteins are O-linked and are often very high molecular weight molecules (species of more than 1 million daltons have been observed). Separation of these molecules using conventional electrophoretic techniques is difficult or even impossible. It may be assumed that these molecules stay in the stacking gel of conventional SDS–PAGE gels (see Welsch *et al.*, 1988).

6.2 Sample preparation

6.2.1 Serum and other biological sample glycoproteins

Serum and other biological samples such as urine, ascites and pleural effusions contain glycoproteins already in solution. They are therefore extremely easy to prepare for electrophoretic analysis. All that is required is dilution in sample buffer (see Section 6.4.1) or, if protein concentration is low (e.g. as in urine), concentration by the method given in *Box 6.3* followed by dilution in sample buffer.

We have found, however, that serum and other biological samples often contain a very high concentration of albumin which can significantly distort electrophoretic separations (see *Figure 6.1*). To overcome this problem, albumin can be selectively adsorbed using an ion exchange resin called 'Affi Gel Blue' (Bio-Rad). The method is given in *Box 6.1*.

Box 6.1: Use of Affi Gel Blue to remove albumin from samples

Affi Gel Blue may be purchased from Bio-Rad (see Appendix B for details). Albumin is selectively adsorbed on to the resin. Other components should not interact with it (but see note 1). If a large volume of sample is to be cleaned up, use a small chromatography column packed with Affi Gel Blue.

1. Pass several column volumes of distilled water through the column to wash and equilibrate it.
2. Apply the sample to the column, and allow it to slowly percolate through the column. Collect sample that does not bind to the column.
3. Elute and collect any residual unbound material from the column using a small volume of distilled water.

If a small sample (e.g. <1 ml volume) is to be cleaned up, we have found that a much more straightforward approach is as follows:

(continued)

---

1. Place approximately 0.5 ml Affi Gel Blue in a 1.5 ml Eppendorf tube.
2. Add sample.
3. Mix in an end-over-end mixer for approximately 30 min.
4. Allow beads to settle or spin briefly in an Eppendorf centrifuge.
5. Draw off supernatant.

Notes
1. In our experience, in addition to albumin, Affi Gel Blue may sometimes selectively remove other (unidentified) minor components from biological samples.
2. When using this method, we generally simply discard the Affi Gel Blue after use. However, the Affi Gel Blue may be regenerated by washing in a solution of 0.05 M Tris and 0.2 M NaSCN or KSCN (adjust pH to 8 using concentrated HCl).
3. Several cycles of Affi Gel Blue treatment using a slow flow rate are sometimes required to remove the majority of the albumin from serum.

6.2.2   Tissue-bound glycoproteins

In contrast to soluble glycoprotein samples, membrane glycoproteins from cells and tissues need solubilization. The method given in *Box 6.2* effectively disrupts cells and tissues and releases a wide spectrum of proteins for analysis. The chloroform/methanol protein precipitation method given in *Box 6.3* is extremely useful in preparation of samples for SDS–PAGE. It effectively precipitates the proteins and removes any contaminating lipids.

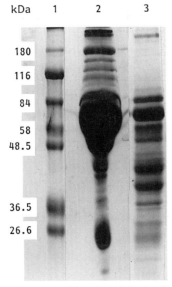

Figure 6.1:   Distortion of electrophoretic separations by albumin.
Albumin can distort electrophoretic separations significantly. It can be effectively removed by treatment with Affi Gel Blue. Track 1, molecular weight markers; track 2, sample showing massive distortion by albumin; track 3, after treatment with Affi Gel Blue albumin component is much reduced.

Box 6.2: Release of glycoproteins from cells and tissues for SDS–PAGE

1. Finely chop a small piece of solid tissue using a clean scalpel blade, then homogenize it in a minimum volume of distilled water or lysis buffer (see Appendix A for recipe) using an electric tissue homogenizer until a fine slurry is formed.
   For cells from cell culture we recommend ultrasonic disruption (sonicate on ice in a minimum volume of distilled water in 5 sec bursts) or, preferably, nitrogen cavitation with a Parr nitrogen bomb (apply 700 p.s.i. for 15 min, then release the pressure).
2. Check the result under the microscope. Some intact cells and lumps of tissue may remain, but most cells should be disrupted.
3. Separate the soluble fraction by centrifugation for 15 min in an Eppendorf centrifuge at 10 000 *g*. Cell debris and intact cells/tissue will pellet while solubilized proteins will remain in solution.
4. Precipitate the proteins in the supernatant by the chloroform/methanol protein precipitation method described in *Box 6.3*.

Box 6.3: Method for concentrating proteins and removing detergents, lipids and salts (after Wessel and Flügge, 1984)

1. Take a 400 µl sample, add 400 µl methanol and 100 µl chloroform.
2. Vortex for approximately 1 min to mix.
3. Centrifuge in an Eppendorf centrifuge at 10 000 g for 2 min: an upper aqueous/methanol phase will be seen, a lower lipid/chloroform phase, and the proteins should be precipitated as a visible disc at the interphase (*Figure 6.2*).
4. Discharge upper methanol/aqueous layer and lower chloroform/lipid phase by suction with a Pasteur pipette.
5. Add 200 µl methanol (the protein yield from several test tubes may be pooled in this step). Vortex briefly. Centrifuge in an Eppendorf centrifuge at 10 000 *g* for 2 min. The protein will pellet to the bottom of the tube.
6. Suck off all liquid and allow the pellet to dry for 30 min in the test tube.
7. To run on SDS–PAGE gels, dissolve the protein in a minimum volume of 10% (w/v) SDS solution in distilled water overnight.

Transmembrane glycoproteins, which must have hydrophobic regions in order to span the membrane, pose particular problems and can be most elegantly solubilized and separated from other proteins

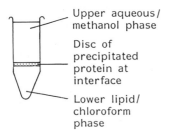

Upper aqueous/methanol phase

Disc of precipitated protein at interface

Lower lipid/chloroform phase

Figure 6.2: The chloroform/methanol precipitation method for concentrating proteins and removing detergents, lipids and salts.

by the use of the detergent Triton X-114 using the method given in *Box 6.4*. At 20°C, Triton X-114 behaves like a lipid, while at 0–4°C it is water soluble. In the transition between the two temperatures, membrane proteins are incorporated into the developing micelles. Afterwards the detergent can be very effectively removed by the chloroform/methanol precipitation method given in *Box 6.3* (see Trudrung and Schumacher, 1989).

Box 6.4: Separation of membrane glycoproteins by Triton X-114

1. Homogenize minced tissues or cells in a 2% Triton X-114 solution in 0.1 M PBS in a tissue homogenizer at 4°C. Incubate with the detergent for 1 h.
2. Centrifuge at 4°C (bench-top centrifuge, 2500 *g*, swinging bucket rotor, 30 min).
3. Withdraw the supernatant, warm it to room temperature and centrifuge again (20°C, 1000 *g*, 5 min).
4. Take the pellet and add Triton X-114 to the supernatant [2% (v/v) final concentration]. Repeat incubation and centrifugation; pool the two pellets.
5. Remove the detergent and precipitate the proteins by the chloroform/methanol precipitation method given in *Box 6.3*.

6.3 Preliminary analysis of lectin binding by dot-blotting

To check whether a particular lectin will bind to a defined glycoconjugate or a glycoconjugate mixture, dot blotting is a quick, simple and efficient method. It can be conveniently used to screen a large number of samples, or the binding of a large number of lectins to a single sample.

The principle of this technique is that the glycoconjugates are first trapped on to a suitable membrane. Immobilon-N (Millipore) is suitable for glycosaminoglycans and Immobilon-P (Millipore) or nitrocellulose (Schleicher & Schuell) is suitable for glycoproteins. The binding of a lectin to the sample is tested by a simple direct or indirect staining technique. The most convenient label for visualization of lectin binding is, in our experience, horseradish peroxidase with the chromogenic substrate DAB–hydrogen peroxide, as the deep brown reaction product shows up very clearly against the white membrane.

A method for dot blotting to test for lectin binding to glycosaminoglycans and glycoproteins is given in *Box 6.5*.

Box 6.5: Dot blot assay to detect lectin binding to glycosaminoglycans and glycoproteins (after Schumacher *et al.*, 1992)

1. For glycosaminoglycans, soak strips of Immobilon-N membranes (Millipore) in 95% (v/v) ethanol in distilled water for a few seconds and then wash extensively in distilled water. For glycoproteins, soak strips of Immobilon-P membranes (Millipore) in 95% (v/v) ethanol in distilled water for a few seconds and then wash extensively in distilled water or use nitrocellulose (Schleicher & Schuell) dry, straight from the packet.
2. Using a Hamilton syringe, apply 5–10 µl spots of glycoprotein or glyco-saminoglycans solution [approximate concentration 1 mg ml^{-1} (w/v) in distilled water] direct to the membrane. Allow to air dry.
3. Incubate the blot strips with 1% (w/v) BSA in lectin buffer containing 0.05% (v/v) Tween 20 for 1 h on a shaking platform.
4. Transfer to a 10 µg ml^{-1} solution of peroxidase-labelled lectin in 4% (w/v) BSA in lectin buffer with 0.05% (v/v) Tween 20. Incubate on a shaking platform for 1 h.
5. Wash 3× in 1% (w/v) BSA in lectin buffer.
6. Incubate with DAB–hydrogen peroxide (see *Box 3.11*) for a few minutes until dots show up deep brown against a clean background (see note 1).
7. Wash extensively in running tap water then blot dry between sheets of clean tissue paper (see note 2).

Notes
1. The reaction should be stopped as soon as the background begins to colour. A slight background staining when the membrane is wet is not a problem, as it will fade considerably upon drying.
2. The dot blot will fade and discolour on exposure to light, so should be stored in the dark (e.g. between the pages of a laboratory notebook).
3. Always use positive samples as controls (e.g. serum glycoproteins).
4. Glycosaminoglycans may be stained by an 1% (w/v) Alcian blue solution in 4% (v/v) acetic acid in distilled water (pH 2.5) for non-specific detection.
5. Glycoproteins may be stained in Ponceau Red S solution for non-specific detection using the method given in *Box 6.14*.

6.4 Analysis of glycoproteins by SDS–PAGE and Western blotting using lectins

---

The following SDS–PAGE procedures are adopted for the use of mini gel apparatus such as that illustrated in *Figure 6.3*. These offer the advantage over conventional full-size gel SDS–PAGE that minimal amounts of sample and reagents are needed and the resolution of bands in the 20–120 kDa molecular weight range is excellent.

6.4.1 Preparing and running the gels

All recipes required for SDS–PAGE separation of proteins are given in Appendix A, and the protocols are given in *Boxes 6.6–6.9* and *Table 6.1*.

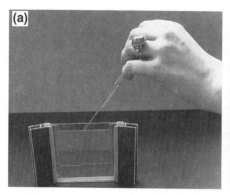 

Figure 6.3:   Typical mini gel apparatus.
(a) Gel casting. (b) Running a gel.

Box 6.6: Preparation of gels

Assembly of glass plates.
1. Clean the glass plates carefully with a normal dishwashing detergent, rinse in distilled water and wipe dry.
2. Assemble the glass plates with the spacers between them (for blotting procedures, thin gels are of advantage over thick gels, so use 0.75 or 0.5 mm spacers).
3. Check if the assembly is leak proof by filling distilled water between the glass plates up to the upper rim. Let stand for about 10 min and see whether leakage has occurred. Pour the water away before adding the gel forming solutions.

Preparation of the separating gel .
1. Mix separating gel buffer, acrylamide stock solution and water in the proportions listed in *Table 6.1*.
2. To polymerize the gel, swirl ammonium persulphate (APS) and TEMED into the mixture. Quickly pipette it between the glass gel plates to a height of about 2 cm from the top of the plates.
3. Pipette approximately 1 ml of water very gently into place above the polymerizing gel to ensure that the upper edge is a completely flat surface.
   NB. Polymerization begins to take place within about 5 min of addition of the APS and TEMED and should be complete within 30 min–1 h.
4. Once polymerization is complete, discard the layer of water. The gel is now ready for addition of the stacking gel.

Notes
1. Higher percentage (e.g. 12.5% or 15%) acrylamide gels will give better resolution of lower molecular weight species, and will give poorer resolution of high molecular weight proteins (which may not even enter the gel). Gels of 15% are really quite stiff and may be difficult to handle because they are fairly brittle and can break.
2. Lower percentage (5–7.5% acrylamide) gels are better for resolving higher molecular weight proteins. They are however very soft and difficult to handle.
3. For most purposes, or when beginning with an uncharacterized sample, we recommend starting with a 10% gel.
4. Prepare the APS solution fresh daily: 100 mg in 1 ml distilled water. If polymerization takes longer than 30 min, you can increase the amount of APS added.

(*continued*)

Preparation of the stacking gel.

1. Mix together 1.25 ml of stacking gel buffer, 0.5 ml acrylamide stock and 3.25 ml distilled water.
2. To polymerize, quickly swirl in 30 μl APS (see note 1) and 10 μl TEMED.
3. Pipette quickly between the plates on top of the separating gel.
4. Quickly insert the sample comb (see note 1).
5. Allow to polymerise (note 2) for approximately 30 minutes.

Notes

1. When inserting the sample comb, avoid trapping air bubbles underneath it. The best way to do this is to insert one corner first then lower the rest of the comb gently into place.
2. Polymerization will commence very quickly (within 1 or 2 min) so work fast. Polymerization should be complete within 15–30 min.

Box 6.7: Assembly of the apparatus

1. Remove the clamps from the gel plates and remove the rubber spacers.
2. Place the gels into the electrophoresis chamber and clamp into position.
3. Fill the lower reservoir with running buffer by tilting the whole apparatus to avoid air bubbles under the lower rim of the glass plates. While gradually returning the apparatus to the horizontal position, slowly add more buffer (see note 1).
4. Fill the upper buffer reservoir until the entire gel is submerged.
5. Gently remove the sample comb so that buffer flows into the sample wells (see note 2).

Notes

1. A carefully directed jet of buffer will displace any trapped air bubbles: use a 20 ml syringe fitted with a hypodermic needle.
2. If the sample wells become crooked or distorted as the comb is removed, they may be easily straightened by gently probing with a Hamilton syringe.

Box 6.8: Sample preparation

1. Mix the protein sample with either reducing or non-reducing sample buffer (note 1) in a 1:1 ratio in an Eppendorf tube.
2. For reducing conditions, pierce the cap of the tube, and heat in a boiling water bath for 1–2 min. Allow to cool.
3. Apply 25–30 μl of sample to each well using a Hamilton syringe (see notes 2 and 3).

Notes

1. It is most usual to separate the sample molecules under reducing conditions (i.e. incorporating reducing sample buffer) as this gives optimum resolution of bands. Samples are diluted 1:1 in reducing sample buffer and heated briefly in a boiling water bath (1–2 min). Disulphide bonds are broken and the protein/glycoprotein is reduced to single chain polypeptides. Sometimes, it may be desirable to run gels under non-reducing conditions [e.g. where a sample has already been run under reducing conditions and one wants to know whether a glycoprotein band is linked by disulphide bonds to another (glyco)protein chain]. Here the sample is diluted 1:1 with non-reducing sample buffer and the heating step is omitted.
2. The sample buffer contains a high concentration of glycerol or sucrose which makes it much denser than the running buffer. If the Hamilton syringe is lowered to the bottom of the sample well and the sample gently applied, it will fill the well from the bottom upwards, and will lie in the sample well without spilling over or diffusing into the buffer.

(continued)

The deep blue colour of the sample buffer (provided by the incorporation of Bromophenol blue) means that it is very easy to monitor loading of the sample on to the gel visually.

3. For most membrane preparations about 10 μg of protein per sample well, using a 10-teeth comb, in a 0.5 mm thick gel, is a good starting point. Protein concentration can be assayed by a commercially available kit (e.g. protein determination kit from Bio-Rad), however, serial dilutions of sample may initially give some idea of separation success.

Box 6.9: Running the gel

1. Place the lid on the electrophoresis chamber.
2. Connect the wires of the chamber with the power supply, use 50–100 V initially for 30 min, then 150 V until the Bromophenol blue marker dye has reached the end of the gel (see note 1).
3. After termination of the run, decrease the voltage to zero and switch off the power pack; remove the gel plates from the apparatus and release the gel from between them (see note 2). Gels may then either be stained with CBB (*Box 6.10*) or blotted for probing for lectin binding (*Box 6.15*).

Notes

1. Voltage and times given are a rough guide: you may decrease voltage to extend running time or increase it to speed up the running time in order to adjust the procedure to your time schedule. In our hands, a long run at 50 V gives enhanced resolution of glycoprotein bands. If the gel is run too quickly, heating effects may denature proteins and cause the gel to crack. Very high voltage will cause such a degree of heating that the glass gel plates themselves may crack or shatter.
2. The easiest way to remove the gel from between the two glass plates is as follows: lie the plates flat on the bench, insert a spatula between the plates in the centre of one side of the gel and gently prise the plates apart. Do not prise the plates apart at a corner, or the plate is likely to crack or chip. The gel may then be lifted free of the plates using dampened gloved fingers.

Table 6.1: Recipes for SDS–polyacrylamide separating gels

	Concentration of acrylamide in the gel					
	5%	6%	7.5%	10%	12.5%	15%
Separating gel buffer (ml)	2.5	2.5	2.5	2.5	2.5	2.5
Acrylamide stock (ml)	1.65	2.0	2.5	3.35	4.2	5.0
Distilled water (ml)	5.85	5.5	5.0	4.15	3.3	2.5
To polymerize the gel						
APS (see *Box 6.6*, note 4) (μl)	60	60	60	60	60	60
TEMED (μl)	5	5	5	5	5	5

6.4.2 Molecular weight markers

It is usual to run a series of molecular weight markers in parallel with test samples. These have two main purposes: they may be used as a rough guide to the molecular weight of bands of interest or to accurately calculate the molecular weight of specific protein/ glycoprotein bands, and also their satisfactory resolution provides an indication that the electrophoretic run has been generally successful.

A number of molecular weight standard kits are available commercially (for example, from Sigma). They contain standard proteins or glycoproteins of a range of molecular weights. The experimenter may purchase a ready-made kit, or buy marker proteins separately and create his or her own molecular weight standards. In our experience, kits are convenient and fairly economical and we would recommend their use. In particular, pre-stained molecular weight marker kits are available from a number of commercial sources. These are extremely useful in that the progress and success of electrophoretic separation can be monitored during the run, and pre-stained molecular markers will also be visible on a subsequent Western blot. Examples of some common molecular weight markers are given in *Table 6.2*.

Molecular weight markers should be chosen to be appropriate for the glycoproteins the experimenter is interested in. For example, if

Table 6.2: Examples of commonly used molecular weight markers for SDS–PAGE[a]

Broad molecular weight range (27.0–180.0 kDa)	
Triosephosphate isomerase	26.6
Lactic dehydrogenase	36.5
Fumarase	48.5
Pyruvate kinase	58.0
Fructose-6-phosphate kinase	84.0
β-galactosidase	116.0
α-2 macroglobulin	180.0
Low molecular weight range (14.0–70.0 kDa)	
α-lactalbumin	14.2
Trypsin inhibitor	20.1
Trypsinogen	24.0
Carbonic anhydrase	29.0
Glyceraldehyde-3-phosphate dehydrogenase	36.0
Ovalbumin	45.0
BSA	66.0
High molecular weight range (29.0–205.0 kDa)	
Carbonic anhydrase	29.0
Ovalbumin	45.0
BSA	66.0
β-galactosidase	116.0
Myosin	205.0

[a] All are available individually or as a kit from Sigma.

you are interested in a series of large glycoproteins in the molecular weight range 90–120 kDa, a high molecular weight and not a low molecular weight kit is most suitable.

6.4.3 Non-specific visualization of proteins and glycoproteins

After separation in the slab gel, the proteins must then be visualized. CBB is a simple and reliable stain (*Box 6.10*). Additionally, sugar residues on glycoproteins may be visualized by the periodic acid–Schiff (PAS) reaction (*Box 6.11*). However, it is only indicative if a protein is highly glycosylated. Alternatively, commercially available kits may be used as general carbohydrate indicators (e.g. glycan kit, Boehringer Mannheim).

Box 6.10: CBB total protein stain for SDS–PAGE gels

Recipes for CBB staining solution and destaining solution are given in Appendix A.

1. Transfer the gel to a dish of CBB staining solution. Allow to incubate on a shaking platform for 1–2 h.
2. Transfer the gel to a dish of destaining solution. Allow to incubate on a shaking platform until the destaining solution becomes discoloured. Pour away the discoloured destaining solution (see note 1) and replace with fresh destaining solution.
3. Incubate on a shaking platform in several further changes of destaining solution until protein bands show up navy blue against a clean, transparent background.

An example of a CBB-stained gel is given in *Figure 6.4*.

Notes
1. The destaining solution can be recycled by passing through a funnel containing a filter of activated charcoal.
2. When the destaining procedure is complete, the gel may be photographed to form a permanent record, or may be stored in 9% (v/v) acetic acid in distilled water for several weeks or months, or dried in a gel dryer.
3. This stain has the advantage that it is stoichiometric and density scans can be used to calculate the amount of protein in a band.

Box 6.11: Periodic acid–Schiff (PAS) reaction to demonstrate glycoproteins in SDS–PAGE gels

1. Fix gels for 30 min in 12.5% TCA in distilled water.
2. Wash 2× for 5 min in 10% acetic acid.
3. Transfer to freshly prepared 1% (w/v) periodic acid in distilled water for 30 min.
4. Wash for 3× for 10 min in 10% acetic acid.
5. Transfer to Schiff's reagent for 30 min.
6. Wash in distilled water until background is clear.
7. Photograph immediately for a permanent record – PAS stain fades quickly.

(*continued*)

Troubleshooting

1. PAS stain gives unimpressive results. We have often experienced poor results with the PAS stain. It is only appropriate for very heavily glycosylated molecules which may not form a large component of any given sample. If the PAS stain shows few bands, we would recommend using the CBB stain (an excellent and virtually foolproof general protein stain) instead. Gels may be successfully stained with CBB after the PAS reaction has been carried out.

2. CBB stained bands are faint. There may be two reasons for this:
   - Protein loading is insufficient. Try concentrating your sample (e.g. using the methanol–chloroform precipitation method given in *Box 6.3*).
   - CBB stain has lost its activity. Make a fresh solution and try again (gels can be re-stained without any problems).

3. Bands are 'wiggly'. 'Wiggly', distorted bands usually indicate contamination of the sample with, for example, salts. Clean up the sample using the methanol–chloroform precipitation method given in *Box 6.3* and try again.

4. Bands are smeared.
   - Smearing of bands is often due to lipid contamination of sample. It is frequently a problem when fatty tissue samples (e.g. breast tissue) are analysed. Clean up the sample using the methanol–chloroform precipitation method given in *Box 6.3* and try again.
   - Smearing may also be caused by solid debris in the sample slowly solubilizing during the run and being released into the gel. Centrifuge samples in an Eppendorf centrifuge at 10 000 *g* for 10 min before use.

5. Distorted lanes.
   - A common cause of distorted lanes is the trapping of air bubbles underneath the gel plates during the run – the voltage will not pass through an air pocket. Ensure at the beginning of the run that no air bubbles are trapped underneath the gel plates (the correct way in which to fill the lower chamber with running buffer is given in *Box 6.7*), and check at intervals during the run that no air bubbles form (if they do, stop the run and release them by gently tilting the chamber). It should be possible to monitor the progress of the run by observing the Bromophenol blue marker dye (it should proceed in a completely straight line). Any 'kinks' may indicate trapped air bubbles.
   - A second cause of distorted lanes may be uneven protein loading of samples running in adjacent lanes. If one lane is overloaded, the protein bands will appear to bulge out and encroach on the adjacent lane. Either take care that samples contain comparable, and appropriate, protein concentrations, or leave an empty lane between samples.
   - Many biological samples contain a very high concentration of albumin. This can cause a great deal of distortion in the run. Remove excess albumin using the simple method given in *Box 6.1*.

6.4.4 Western blotting

SDS–PAGE gels cannot be directly probed for lectin binding to glycoproteins. In order to investigate lectin binding, separated proteins must first be transferred to a membrane, such as nitrocellulose, using the technique of Western (protein) blotting.

Western blotting may be carried out in two ways, either wet (tank) electroblotting (see *Box 6.12*) or semi-dry electroblotting (see *Box 6.13*). Both procedures give equally good results, and the choice of method depends entirely on the worker's individual preference.

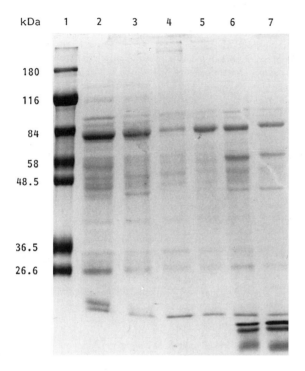

Figure 6.4:   An example of a CBB stained polyacrylamide gel.
Track 1, molecular weight markers; tracks 2–7, breast cancer glycoproteins.

Generally speaking, wet blotting takes longer (at least 1 h) and uses more transfer buffer, but gives quantitative as well as excellent qualitative results. Semi-dry blotting is quicker (typically 15–30 min), uses very small volumes of transfer buffers and gives excellent qualitative, but not quantitative, results.

Box 6.12: Tank (wet) Western blotting of SDS–PAGE gels

The recipe for transfer buffer is given in Appendix A.

NB. Wear gloves to avoid fingerprints on the nitrocellulose!

Assembly of the blotting apparatus.
1. Cut six sheets of filter paper (note 1) and a sheet of nitrocellulose (note 2) to the size of the gel.
2. Wet the filter paper in transfer buffer (note 3).
3. Lay the gel (note 4) on top of three sheets of wetted filter paper, avoid trapping air bubbles under the gel (note 5).

(continued)

4. Lay nitrocellulose (soaked in transfer buffer) on top, again avoid air bubbles!
5. Cover nitrocellulose with the final three wetted filter papers.
6. Put this sandwich between the pads of the blotting tank cassettes, with the nitrocellulose facing toward the anode and the gel toward the cathode, set up the cooling system and fill the tank with transfer buffer.
7. For a normal sized mini-gel blot for about 1 h at 0.25 A (note 6).
8. When the run is finished assess blotting efficiency by Ponceau Red S staining (see *Box 6.14*).

Notes

1. One can use special electrode filter paper for electrophoresis (Whatman or Schleicher & Schuell). Alternatively, Whatman number 1 filter paper can be used, but as this is three times thinner, 18 sheets are required.
2. We recommend Bio-Rad or Schleicher & Schuell nitrocellulose.
3. Soaking filter papers and nitrocellulose is most conveniently done in large Petri dishes or in plastic sandwich boxes.
4. If unstained molecular weight markers have been used, the part of the gels which contain the molecular weight markers should be cut off and stained with CBB before blotting (*Box 6.10*). Pre-stained molecular weight markers (see Section 6.4.2) are a particularly good idea when the gel is to be blotted: their visible transfer is a good guide that blotting has been successful, and they may be used for calculation of precise molecular weights of any transferred glycoprotein bands of interest.
5. Air bubbles may be expelled using the flat edge of a spatula, or by rolling with a glass rod. We have also found a small rubber roller, purchased from an art shop or photographic supplier, to be useful for this purpose.
6. Small molecular weight proteins are more mobile and will transfer more rapidly than larger molecular weight proteins. Blot at 0.25 A for 1 h per gel as a good general guide; however, if you are interested in low molecular weight species, you may wish to blot for a shorter time; similarly, higher molecular weight proteins will be optimally blotted after a longer blotting period. Experiment!

Box 6.13: Semi-dry Western blotting of SDS–PAGE gels

Recipes for anode buffers I and II and cathode buffer are given in Appendix A.

1. Cut six sheets of filter paper (note 1) and a sheet of nitrocellulose (note 2) the same size as the gel.
2. Soak one sheet of filter paper in anode buffer I; two sheets in anode buffer II and three sheets in cathode buffer. Soak the nitrocellulose in distilled water (note 3).
3. Lay the sheet of filter paper soaked in anode buffer I directly on to the centre of the anode; lay the two sheets of paper soaked in anode buffer II on top; and the nitrocellulose on top of that.
4. Smooth out any air bubbles (note 4).
5. Carefully lay the gel on top (note 5).
6. Cover with the three sheets of filter paper soaked in cathode buffer, and smooth out any air bubbles.
7. Put the cathode in place.
8. Connect up to a power pack. Blot at a current of 0.25 A per mini-gel for 30 min (see notes 6 , 7 and 8).

The 'sandwich' of gel, nitrocellulose and filter papers is illustrated in *Figure 6.5*.

(*continued*)

Notes

1. One can use special electrode filter paper for electrophoresis (Whatman or Schleicher & Schuell). Alternatively, Whatman no. 1 filter paper can be used but, as this is three times thinner, 18 sheets are required.
2. We recommend Bio-Rad or Schleicher & Schuell nitrocellulose.
3. Soaking filter papers and nitrocellulose is most conveniently done in large petri dishes or in plastic sandwich boxes.
4. This can be done with the flat edge of a spatula or by rolling with a glass rod. We have found that an excellent accessory is a small screen printing roller purchased from an art supplier. Roll very gently so as not to distort the gel.
5. If unstained molecular weight markers are used, the part of the gels which contains the molecular weight markers should be cut off before blotting and stained with CBB (*Box 6.10*). Pre-stained molecular weight markers (see Section 6.4.2) are a particularly good idea when the gel is to be blotted: their visible transfer is a good guide that blotting has been successful, and they may be used for calculation of precise molecular weights of any transferred glycoprotein bands of interest.
6. If more than one gel is to be blotted simultaneously, increase the current (i.e. 50 mA for 30 min for two gels; 75 mA for 30 min for three gels) or increase the blotting time (i.e. 0.25 A for 1 h for two gels; 0.25 A for 1.5 h for three gels, etc.). If the current is increased, beware excessive heating. See also note 7.
7. It is a good idea to keep a wary eye on the current during blotting. We have found that it sometimes fluctuates quite dramatically during the first 5 min of blotting and may need readjusting. After that, it should be checked every few minutes. If the current falls, blotting will be less efficient; if the current increases, blotting will occur faster and proteins may pass straight through the nitrocellulose and on to the cathode filter papers! Higher current may also lead to excessive heating: expect the blotter to feel slightly warm to the touch after blotting, but avoid any larger temperature rise as proteins will be denatured. If excessive heating occurs, the gel may char and the blotter may be damaged.
8. Small molecular weight proteins are more mobile and will transfer more rapidly than larger molecular weight ones. Use 0.25 A for 30 min per gel as a good general guide for protein blotting, however, if you are interested in low molecular weight glycoproteins, you may wish to blot for a shorter time; similarly, higher molecular weight proteins will be optimally blotted after a longer blotting period. Experiment!

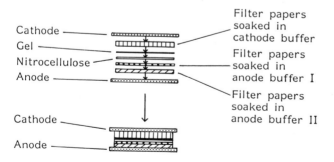

Figure 6.5: Semi-dry electroblotting.

The arrangement of polyacrylamide gel, nitrocellulose membrane, filter paper and electrodes.

After transfer of the proteins on to nitrocellulose by Western blotting, which can be done either by tank or by semi-dry blotting, efficiency of transfer should be estimated by Ponceau Red S staining (*Box 6.14*) before determination of the lectin binding sites using the method given in *Box 6.15*. An example of results is given in *Figure 6.6*.

Box 6.14: Ponceau Red S total protein stain for blots

1. Immerse the nitrocellulose in 0.1% (w/v) Ponceau Red S solution in 5% (v/v) glacial acetic acid in distilled water for a few seconds.
2. Wash with several changes water until pink bands appear against a white background.

Note

1. The blot may be taken from the distilled water and blotted between sheets of filter paper to give a permanent record. Alternatively it can be completely destained by further washing in distilled water, or may be transferred straight to the blocking solution (*Box 6.15*) for staining for lectin binding (it will completely destain in this solution).

Troubleshooting

1. Poor transfer. Poor transfer implies that conditions were not ideal.
   - The most common reason is that the gel was blotted for too long or too short a period or at a too high or low current: use the guidelines given in the methods above, and experiment to optimize transfer conditions. If high molecular weight proteins appear to be successfully transferred while low molecular weight proteins are not, decrease current or decrease blotting time. If low molecular weight proteins appear to be successfully transferred while high molecular weight proteins are not, increase current or increase blotting time. You can often control for these conditions by using a second layer of nitrocellulose on top of the first one. If the current was too high or blotting too long for the proteins of interest they will pass through the first membrane and be trapped on the second.
   If transfer is generally poor overall, this may imply that:
   - Fresh blotting buffers are required (blotting buffers contain methanol which can evaporate on storage and therefore deteriorate).
   - The blotting 'sandwich' was not set up correctly. Check that the instructions given above have been adhered to, that air bubbles have been excluded, and that anode and cathode are in close contact with the 'sandwich'.
   - There is a mechanical fault with the blotting apparatus. If you are performing tank blotting, check the platinum wires of the blotting chamber.

2. Distorted bands. Check that the gel itself ran efficiently and separated proteins successfully. It is a good idea to run two identical gels simultaneously and stain one with CBB total protein stain (*Box 6.10*) to check if there are problems (see the 'Troubleshooting' section attached to *Box 6.10*).

   If the gel ran well, the most common causes of distorted bands on the blot are:
   - Air bubbles. Take great care during the setting up of the blotting 'sandwich' to exclude all air bubbles. See note 4 of *Box 6.13*. Proteins will not transfer across an air pocket. Air bubbles will result in a spotty or patchy appearance of the blot.
   - Squashing or stretching the gel. The gel should be handled with great gentleness. Low percentage acrylamide gels are especially fragile and susceptible to stretching and tearing. Do not be too vigorous when excluding air bubbles from the blotting 'sandwich'. Do not squash the gel.

Box 6.15: Staining for lectin binding sites on Western blots (*Figure 6.6*)

1. After Ponceau Red S staining, transfer the blot to a dish of blocking buffer [4% (w/v) BSA in 0.05% (v/v) Tween 20 in lectin buffer; stable at 4°C for up to a month] on a shaking platform for 30 min.
2. Transfer the blot to a dish of horseradish peroxidase-labelled lectin at a concentration of 1–10 μg ml^{-1} in 0.05% Tween 20 in lectin buffer. Incubate on a shaking platform for 1 h (see notes 3 and 4).
3. Wash the blot 3× for 5 min in 0.2% Tween 20 in lectin buffer.
4. Incubate with DAB–hydrogen peroxide (*Box 3.11*) for 5–30 min until bands show up distinct and dark against a a clean background (see note 5).
5. Wash extensively in running tap water.
6. Blot dry between clean tissue paper (see note 6).

Notes
1. Controls for the specificity of the lectin reaction can be performed by incubation of the appropriate inhibitory sugars with the blot (e.g. incubation with the labelled lectin at 10 μg ml^{-1} in 0.05% Tween 20 in lectin buffer with 0.1 M appropriate monosaccharide added should effectively inhibit most specific binding). Serial dilutions of inhibitory sugars may produce useful information about those glycoproteins which bind the lectin most tightly (e.g. protein band 'a' does not bind the lectin in the presence of 0.1 M sugar, while band 'b' still does).
2. Indirect methods (e.g. using biotin-labelled lectin followed by streptavidin peroxidase) may be used to detect lectin binding to glycoproteins immobilized on nitrocellulose, and these may offer advantages over the direct method as the large horseradish peroxidase label may interfere with the lectin binding site in some cases (see Section 4.2).
3. Generally speaking, lower concentrations and longer incubation times work on blots compared to tissue sections in cell and tissue histochemistry (Chapter 4) as the blot is far more sensitive. Investigators are encouraged to experiment and to optimize conditions for any individual system.
4. We have found that it is often convenient to run a gel and carry out blotting on one day, then to incubate the blot with labelled lectin at low concentration (e.g. 1 μg ml^{-1}) overnight, and to develop the blot on the second day. As long as the blot is not allowed to dry out and is kept moving on a shaking platform, this poses no problem. Alternatively, blots may be incubated with blocking buffer on a shaking platform for 30 min, then dried and stored for a few days before lectin binding is carried out.
5. We have observed that, even if the background comes up during the DAB–hydrogen peroxide stage, low to moderate background staining will fade to nothing when the blot dries.
6. Once the blot is dry, photograph within a few days for a permanent record as results fade gradually over time. Store blot in the dark (e.g. sandwiched in a laboratory notebook) as it will discolour if exposed to light for prolonged periods.

Troubleshooting
1. Lectin binding poor. This may be due simply to the fact that there is a very low concentration of glycoprotein(s) recognized by that lectin actually present on the blot. It is a good idea to carry out a quick dot blot (*Box 6.5*) of your sample in parallel with a positive control, such as the appropriate monosaccharide linked to a carrier protein like BSA (a wide range is available from Sigma) and stain that for lectin binding using the method given in *Box 6.15*. If staining of the control is strong and staining of the sample is weak on the dot blot, where glycoproteins are concentrated into a tiny spot, this indicates that there is simply a very low concentration of the lectin-binding glycoprotein in your sample.

(continued)

If staining of the control is poor, this indicates that the lectin binding method is at fault – try increasing lectin concentration, or extending incubation time, or adopting a more sensitive method (e.g. if you are using peroxidase-conjugated lectin, try biotinylated lectin followed by streptavidin peroxidase (see section 4.5.1), or an ABC technique (see section 4.5.3). It is also worth checking that reagents - lectin, DAB, buffers etc. haven't deteriorated.

2. High background staining
   - Ensure that 4% (w/v) BSA and 0.05% (v/v) Tween 20 are incorporated into all incubation and washing solutions. Heating the blocking solution to 56°C for 30 min may also reduce background staining.
   - Ensure that all incubations take place on a shaking platform.
   - Ensure that the blot does not dry out at any point during the staining protocol.

6.5 Analysis of glycoproteins by IEF and lectin blotting

An alternative method of protein separation is by the use of IEF, which can conveniently be applied directly to tissue sections [it is then

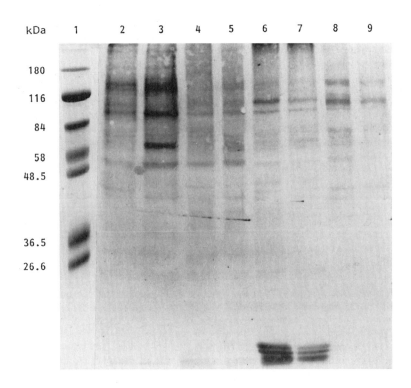

Figure 6.6: Western blot of glycoproteins separated by SDS–PAGE and stained for the binding of HPA.
Track 1, molecular weight markers; tracks 2–9, breast cancer glycoproteins.

referred to as direct tissue isoelectric focusing (DTIF); see Schumacher *et al.*, 1990]. This method has also been adopted for use with a mini IEF chamber from Bio-Rad, which has the advantage of being cheaper to install and run than conventional IEF (Schumacher and Trudrung, 1991). Since no pretreatment of the tissue is necessary and the residues of the tissue sections after IEF are still visible, this approach offers the unique feature that lectin binding to the residual tissue specimen as well as eluted components may be achieved. This method has been successful in our hands. Full methods are given in *Boxes 6.16–6.18.*

Box 6.16: Direct tissue isoelectric focusing (DTIF)

Since cryostat sections can be placed directly on to an IEF gel, no special sample preparation is required. Recipes are given in Appendix A.

Casting of the gels (mini-IEF chamber).
1. Moisten two glass plates the size of the PAGE foils (LKB/Pharmacia) with water. Mount one PAGE foil on the glass plates (avoid air bubbles) hydrophilic side upwards, the second hydrophilic side downwards. Tighten the attachment of the foils to the glass plates with a little roller (obtainable from suppliers of photographic equipment or art shops).
2. Use two layers of Parafilm strips (0.5 cm broad) as spacers (0.24 mm thick) and lay them around one glass plate in a U-shaped fashion.
3. Put a cannula attached to a 10 ml syringe in the middle of the open side of the U on one glass plate, put the other one on top of it and clamp the plates together with bulldog clips. Position the gel cassette upright.
4. Prepare the gel solution in a conical flask with a side arm and well-fitting rubber stopper. Dissolve 1 g sorbital in 10 ml of the following solution: 1.7 ml acrylamide/Bis solution, 0.5 ml Servalyte, 8.2 ml distilled H_2O.
5. Stopper the flask, attach a rubber tube to the side arm of the flask, and connect it up to a vacuum pump. De-gas by suction for 5-10 minutes. Release the vacuum, and unstopper the flask.
6. Mix 150 µl APS (see note 1) and 15 µl TEMED into the gel mixture. Suck up into a 10ml syringe and inject into the gel cast. Stop injection when the gel solution is 2 cm away from the top. Remove canula. Polymerization takes about one hour. It is convenient to use the gel next day, store in the refridgerator or cold room, wrapped in clingfilm. Before use, cut the gels to the right size for the mini IEF chamber. Soak graphite electrodes overnight in anode or cathode buffer, respectively.
7. With the aid of some dops of water, mount the gel (PAGE foil downwards) on to the glass plate of the chamber. Absorb any excess water using filter paper.
8. Apply 10-30 µm thick cryostat sections about 1.5 cm away from the margin of the gel.
9. Place the gel on the electrodes (upside down!), close the lid and apply a cooling cushion, pre-chilled in the deep freeze, on top of the lid.
10. Apply current: 100 V for 15 minutes, then 200 V for 30 minutes, then 450 V for 5 hours.
11. After termination of the run stain in CBB (*Box 6.17*) or blot (*Box 6.18*).

Note
1. APS solution: 100 mg APS in 1 ml distilled water. Make fresh daily.

Box 6.17: CBB total protein stain for IEF gels

The times needed for staining vary according to gel thickness; times indicated are for 0.2 mm thick gels. Recipes are given in Appendix A.

1. Fix samples in a 20% (w/v) TCA solution in distilled water for 5 min.
2. Rinse 2× for 2 min in distilled water.
3. Transfer to CBB staining solution for 10 min.
4. Transfer to destaining solution 3× for 10 min.
5. Transfer to impregnation solution for 2 min.
6. Air dry.

Box 6.18: Blotting IEF gels

The principle arrangement is the same as for SDS–PAGE blotting, however, there are some modifications (blotting buffer, loosening of the gels and electrophoretic conditions). Recipes are given in Appendix A.
Blotted IEF gels may be stained for total protein or lectin binding exactly as described in *Boxes 6.14* and *6.15*.

1. Ultrathin gels readily separate from the support foil by the use of a simple device consisting of half of a 15 cm diameter plastic sewage pipe. Place the gel with the wetted nitrocellulose (notes 1 and 2) on top of it on to the pipe and drag a wire between the gel and the support foil. The nitrocellulose can then be removed with the gel laying on top of it.
2. Cut six sheets of filter paper (note 3) to the size of the gel.
3. Moisten three sheets of filter paper in blotting buffer [0.7% (v/v) glacial acetic acid in distilled water (see note 4)].
4. Lay the gel/nitrocellulose gel side downwards on top of the filter paper. Avoid trapping air bubbles under the gel.
5. Cover nitrocellulose with the final three wetted filter papers.
6. Put this sandwich between the pads of the blotting tank cassettes, with the nitrocellulose facing towards the anode and the gel toward the cathode, set up the cooling system and fill the tank with transfer buffer.
7. Blot at 55 V for 40 min.

Notes
1. Before blotting, the part of the gel which contains the markers for the isoelectric point, together with one reference lane, should be cut off and stained with CBB (*Box 6.10*).
2. We recommend Bio-Rad or Schleicher & Schuell nitrocellulose.
3. One can use special electrode filter paper for electrophoresis from Whatman or Schleicher & Schuell. Alternatively, Whatman no. 1 filter paper can be used but, as this is three times thinner, 18 sheets are required.
4. Soaking filter papers and nitrocellulose is most conveniently done in large Petri dishes or in plastic sandwich boxes.
5. Tank blotting is preferred if urea is added to enhance the resolution of the gel. The urea will readily wash out in the large volume of blotting buffer, thus reducing the denaturing effects of the urea.

References

Schumacher U, Trudung P. (1991) Direct tissue isoelectric focussing on mini ultrathin polyacrylamide gels followed by subsequent western blotting, enzyme detection, and lectin labelling as a tool for enzyme characterisation in histochemistry. *Analyt. Biochem.* **194**, 256–258.

Schumacher U, Trudung P, Ruhnke M, Gossrau R. (1990) Direct tissue isoelectric focussing on ultrathin polyacrylamide gels. applications in enzyme-, lectin- and immunohistochemistry. *Histochem. J.* **22**, 433–438.

Schumacher U, Thielke E, Adam E. (1992) A dot blot technique for the analysis of interactions of lectins with glycosaminoglycans. *Histochem. J.* **24**, 453–455.

Trudrung P, Schumacher U. (1989) Analysis of wheat germ agglutinin (WGA)-, *Phaseolus vulgaris* leucoagglutinin (PHA-L)-, and *Lens culinaris* (LCA)-binding to isolated rat neocortical membrane glycoproteins and to brain tissue sections. *Brain Res.* **497**, 399–401.

Welsch U, Buchheim W, Schumacher U, Schinko T, Patton S. (1988) Structural, histochemical and biochemical observations on horse milk fat globule membranes and casein micelles. *Histochemistry* **88**, 357–365.

Wessel D, Flügge IU. (1984) A method for the quantitative recovery of protein in dilute solution in the presence of detergents and lipids. *Ann. Biochem.* **138**, 141–143.

7 Interpretation and Analysis of Lectin Binding

7.1 Interpretation of lectin binding to tissue sections: simple and complex sugars

The approaches and problems of interpreting the significance of lectin binding to cells and extracellular matrix are much the same as those for staining methods involving antibodies and enzymes. It differs in that the assumed monosaccharide specificity (e.g. simply gal, fuc, galNAc, etc.) of a lectin can be very different from the actual complex oligosaccharide structure(s) it recognizes in histochemical preparations or Western blots. As explained in Section 1.5, although we usually refer to a lectin in terms of its monosaccharide specificity (a galNAc- or man-binding lectin, for example), the actual binding partners of lectins in cells and tissues are complex three-dimensional structures, encompassing sub-terminal as well as terminal monosaccharides, with hydrophobic and electrostatic interactions playing an important part in binding specificity. The binding patterns of a panel of 'galNAc-binding' lectins to serial sections of the same tissue, for example, will be astoundingly different (*Figure 3.7*). One of the more common mistakes made when using lectins, is to believe that all 'gal-binding' lectins or all 'fuc-binding' lectins are interchangable and will give similar results. In fact, they will give very different results.

7.2 Lectin binding to a range of binding partners

Lectins vary in their specificity for monosaccharides and monosaccharide linkages to adjacent monosaccharides. Both lectins and oligosaccharides can change shape in solution, allowing them to recognize and bind to different configurations. Some lectins distinguish between anomers of the same structure, some may be specific for man but also accept glc, some prefer to bind gal but galNAc may be acceptable. Such elasticity may be expressed on information sheets of commercial lectins by relating monosacharide competitive inhibitions in descending order, such as man > glc > gal. It is therefore very difficult to predict the likely structures detected by a lectin binding to a tissue section, and oligosaccharides with very different structures can bind equally well to a particular lectin. Further, oligosaccharide structures that bind to an immobilized lectin, such as tissue extract glycoproteins isolated by affinity chromatography, may not be the same as those oligosaccharides in the rigid tissue section that bind to a soluble lectin.

Histochemical attempts to characterize binding partners, such as by enzyme digestion, can introduce a large range of changes and artifacts, and are often expensive, unsuccessful and difficult to interpret (see Section 4.12 for a more detailed discussion).

7.3 Patterns of binding to cells and tissues

7.3.1 Healthy, normal cells and their extracellular matrix

Lectin binding indicates the main sites of glycoproteins, glycolipids and glycosaminoglycans in the cell and its surrounding matrix. It can be increased in active, secretory normal cells, and will often be intensely concentrated at the secretory surface, in secretions, in the Golgi apparatus and between cells (in association with adhesion molecules).

7.3.2 In disease

In diseased tissue, increased lectin binding, particularly in the cytoplasm, is commonly seen. This is most marked in storage diseases; presumably from an accumulation of, and failure to, transport glycosylated molecules. In addition, changes in glycosylation are observed in cancer cells. For example, in normal breast, HPA only stains the luminal surface of the breast duct epithelium: staining is highly localized, intense and very discrete. In about 80% of breast cancers intense staining of a proportion of cancers is seen (Brooks and Leathem, 1991). Here the binding pattern is quite different: there is intense granular staining throughout the cytoplasm with localization around the entire cell surface. This is a very typical finding when comparing lectin binding to normal with malignant cells. It is illustrated in *Figure 7.1*.

7.3.3 Localization of lectin binding

Lectin binding may be observed at a number of different sites at the cell surface and inside the cell in normal and diseased tissue.

Cell surface and extracellular matrix. If one detects lectin binding at the cell surface and in the extracellular matrix, the following structures and molecules may provide the binding partner for the lectin:

1. Glycocalyx of the cell membrane.
2. Secretions.
3. Intercellular junctions.
4. Basement membrane.
5. Extracellular matrix: glycosaminoglycans, collagen, matrix glycoproteins.

Inside cells.
1. Diffuse cytoplasmic (may be non-specific).
2. Granular cytoplasmic (secretory granules).
3. Endosomes, lysosomes and their contents.

(a)   **(b)**

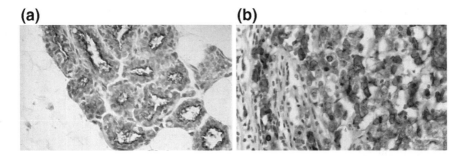

Figure 7.1:   (a) HPA binding to the luminal surface of normal human breast duct. (b) HPA binding to breast cancer, note both cell surface and cytoplasmic staining.

4. Endoplasmic reticulum.
5. Golgi apparatus (especially when located between nucleus and luminal surface).
6. Nucleus/nuclear membranes.

7.4 Recording staining results

7.4.1 'Eyeballing'

There is no right or wrong way of recording staining results of lectin histochemistry. We usually record a rough percentage of cells staining (e.g. if recording lectin binding to cancer cells within a tissue section that contains an inevitably heterogeneous mixture of cancer cells and normal components, we might say that 5%, 10%, 85%, etc., of cancer cells are positive, ignoring everything else. This is just an 'eyeballing' judgement). We also tend to give a rough estimate of staining intensity [e.g. + – (very weak), + , ++, +++, ++++ (very intense)]. A case may therefore be classified as 10% +, or 50% ++ , or 90%++++, etc.

This is, obviously, a relatively subjective judgement, but with experience has proved to be remarkably consistent both in terms of inter- and intra-observer error. It is a simple, quick and convenient way of recording results and one that we have found to be very useful.

7.4.2 How to define positive and negative

Owing to the enormous heterogeneity in glycosylation between cells, and the broad binding specificity of some lectins, it is very unusual to find a cell or tissue preparation that is completely and absolutely negative for binding of any given lectin.

The first thing to consider is what cell populations you are interested in. If you are interested in glycosylation of a particular defined cell type (e.g. endothelial cells, muscle cells, cancer cells, etc.) then *ignore everything else*.

For the record, score the percentage and intensity of staining, as described above, and then decide whether the case is 'positive' or 'negative'. We usually place the cut-off as anything staining less than 5%+ or 50%+– is considered negative, and anything above that is considered positive (Brooks and Leathem, 1991; Leathem and Brooks, 1987). This has proved to be a useful and consistent method of classifying results.

7.4.3 Lectin binding to blood group-specific sugars

As described in Section 1.2, some lectins recognize sugars which are blood group-specific. For example, UEA-I recognizes fuc, which is associated with blood group O/H; HAA recognizes galNAc, which is associated with blood group A, and so on (see *Table 1.1*). Consequently, erythrocytes in a cell or tissue preparation from a blood group A individual will always stain with HAA, those from a group O individual will stain for UEA-I, and so on. In secretors, staining of endothelium and mucins may also be observed (see *Figure 7.2*). This should be considered when interpreting lectin binding results. It should also be borne in mind that in addition to the human blood groups A,B and O there are many other less well known blood group systems in which sugars play an important role (M, N, T, Cad, Tn Tk, etc., see *Table 1.1*), and that animals too have different carbohydrate blood group sugars that can be detected by lectin binding.

7.4.4 Computerized image analysis

There has been much research on image analysis to quantify immunocytochemistry. Although in principle it is an excellent idea (scoring staining by eye having obvious limitations) image analysis carries with it many problems. Field selection is always a problem in image analysis and when reading lectin histochemistry results the problem is magnified, as, owing to the broad distribution of the individual sugars detected by some lectins, their binding to blood group sugars, etc., one must define the fields to be analysed very precisely. This can compromise the objectivity of image analysis techniques very severely. We would advise that one thinks carefully before embarking upon image analysis of lectin binding, and interprets the results with some caution.

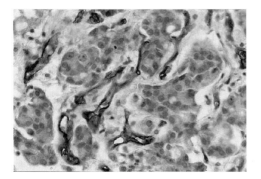

Figure 7.2:   HAA binding to blood vessels in a breast cancer of a blood group A individual. Note: the cancer cells are negative for HAA binding.

7.5 Analysis of lectin binding partners

What can we tell about the carbohydrate structures detected by lectin binding?

7.5.1 Terminal monosaccharides

It is possible to detect the terminal sugar or sugars present . This can be achieved by demonstrating inhibition of lectin binding by one (or more) specific monosaccharide(s), but not by others (see Section 4.7).

It may be possible to determine both terminal and immediately sub-terminal sugars by demonstrating a change in lectin binding after digestion by a terminal glycosidase. As discussed previously (see Section 4.12), this may be more troublesome than it sounds.

7.5.2 More detailed analysis of oligosaccharide structure

Oligosaccharide structures may be released intact from cells or tissues by specific endoglycosidase digestion: enzymes are available commercially that will cleave N-linked or O-linked glycans intact. Alternatively, intact oligosaccharides may be released by temperature-dependent hydrazinolysis (details of these approaches are beyond the scope of this book). The solubilized oligosaccharides may then be characterized in detail by sophisticated techniques such as nuclear magnetic resonance (NMR), fast atom bombardment mass spectrometry (FAB-MS), etc. This area of research is expensive in terms of time and resources, requires specialist equipment and expertise and would be unrealistic for most research laboratories.

There may also be problems of interpretation with these high tech approaches: release of oligosaccharides into solution from their glycoproteins abolishes the spatial arrangement, such as clustering, which may be the keystone of lectin binding specificity in tissues.

References

Brooks SA, Leathem AJC. (1991) Prediction of lymph node involvement in breast cancer by detection of altered glycosylation in the primary tumour. *Lancet* **338**, 71–74.

Leathem AJC, Brooks SA. (1987) Predictive value of lectin binding on breast cancer recurrence and survival. *Lancet* **i**, 1054–1057.

8 Applications and Usefulness of Lectin Binding Studies

Lectins have been widely used as tools for characterization of carbohydrate residues of glycoconjugates in tissue sections and several excellent reviews have been written on this subject (e.g. Damjanov, 1987; Spicer and Schulte, 1992; Walker, 1989). This chapter will focus on the strategy of when and how to apply lectin histochemistry in a research project, and also to give a (necessarily) brief review of some published applications of lectin histochemistry. The reader is advised to consult the many excellent published reviews and databases on this subject for a wider selection.

8.1 Strategy for use of lectin histochemistry

When designing a research project involving lectins, it is worth considering a number of factors, as follows.

8.1.1 Lectins are not species specific

Lectins are carbohydrate specific, and hence are substrate specific *not* species specific. In their binding behaviour they resemble enzymes rather than antibodies. Because of this, the binding behaviour and the affinity of lectins to closely related sugars can be determined in a way similar to the determination of substrate affinity towards enzymes. This has the following practical consequences.

1. Lectins can be more readily applied across species borders than antibodies. While monoclonal antibodies in most instances, and polyclonal antibodies often, do not react across species, lectins

usually do. Lectins react with carbohydrate residues of prokaryotic and eukaryotic cells and their intercellular matrices.

If lectins are applied as a specific marker for a particular cell population rather than as a marker for a particular carbohydrate residue, care must be taken in the choice of lectin(s). For example, UEA-I reacts with human endothelium but not with the endothelium of many other species, including those (rat, mice, rabbit, guinea pig) most widely used in biomedical research. Other lectins may be appropriate for labelling the endothelium of these animals (Alroy et al., 1987). If one is working with unusual species, it is probably worth contacting a laboratory which has a wide range of lectins available, and arranging to screen the suitability of a large panel of lectins for identification of the particular cell type under investigation.

2. Some lectins are blood group specific. The blood group specificity of lectins mainly concerns the ABO blood group system. However, this not only applies to humans but also to some blood group specificities in animals which are also expressed by carbohydrate residues that can be detected by lectins. This has implications for the labelling of cells, since not only erythrocytes but also endothelia and mucin-like glycoproteins can carry blood group antigens and the lectin labelling of cells expressing these structures can vary accordingly. This has to be taken into account when interpreting lectin stained slides.

8.1.2 N-linked and O-linked sugars

According to the linkage of carbohydrate residue with the protein backbone, two main groups of glycoproteins can be distinguished: N-linked and O-linked glycoproteins. The distinction between the two types of glycoproteins has practical consequences: the glycosylation of N-linked glycoproteins starts in the endoplasmic reticulum and the terminal carbohydrate residues are modified and extended within the Golgi apparatus, while the carbohydrate residues of O-linked glycoproteins are entirely added in the Golgi. O-linked glycoproteins are often called mucin-type glycoproteins because this type of linkage is typical of mucins. However, mucins, which are very high molecular weight glycoproteins, are mostly (though not entirely) secreted by epithelial cells and they can also contain N-linked carbohydrate residues (for review see Strous and Dekker, 1992). However, the following rules of thumb apply for the choice of lectins (examples in brackets) to be used in a study.

1. If you are looking predominantly at glycosylation of O-linked glycoproteins, use lectins which react with the following carbohydrate specificities: galNAc (HPA, HAA, DBA), glcNAc

(WGA), gal (RCA, MPA), fuc (UEA-I), and neuraminic (sialic) acid (SNA, MAA). It is of interest to note that some mucins can carry blood group antigens if the carrier of the blood group antigen is a secretor.

2. If you are looking for N-linked glycans, you should use lectins which recognize the mannose core of the N-glycans (GNA, Con A) or those which react with the complex type of carbohydrate residues (PHA-L, PHA-E).

8.1.3 Effect of fixation and processing

Carbohydrate residues detected by lectins may be influenced by the processing of the tissue but, in general, carbohydrate residues (often referred to as glycotopes) are far less affected by fixation than the epitopes of proteins. However, some variation in lectin reactivity can be attributed to fixation. As a rule, unfixed cryostat sections give best results, although tissues fixed in neutral-buffered formalin or formal saline can be used adequately. Often Bouin's fixative gives better results than formalin fixatives (particularly for mucins), but if other staining methods are to be used on the same material it is often wise to use formalin fixation to which most staining methods are adaptable. See Section 3.4.4 for a discussion of fixatives. Trypsin digestion or microwave treatment can be used to enhance lectin reactivity. The incubation time for trypsin treatment or microwaving has to be tested beforehand (see Section 3.4.4): too long a trypsinization/microwaving time can reduce the staining and damage the tissues. As a general rule, the length of trypsinization time required to reveal sequestered carbohydrate residues reflects the length of time for which the tissues were originally in fixative: brief fixation usually requires little or no trypsinization; long fixation times often mean that the specimen will require extensive trypsinization. In our experience, a relatively short trypsin digestion can reduce the background staining considerably, as well as enhance specific binding.

8.1.4 Glycolipids

If carbohydrate residues of glycolipids (e.g. in the investigation of storage diseases) are to be investigated, make sure that no organic solvents like xylene, chloroform or acetone are used as these dissolve lipids and they will be lost (see Section 3.4.4).

8.1.5 Extracellular matrix

Lectins often stain components of the extracellular matrix, a phenomenon that is usually interpreted as non-specific background staining. This is not necessarily the case since many glycoproteins are found in the matrix including, for example, fibronectin and tenascin. In addition, collagen and other fibrils are glycosylated and the carbohydrate residues of the glycosaminoglycans (e.g. hyaluronic acid, heparan sulphate, chondroitin sulphate) are also able to bind certain lectins.

8.1.6 Ask a well defined question

Lectin binding is in part still a tool in search of applications, so for histochemistry 'fishing expeditions' are still reasonable. However, many lectin-based projects are either too naïve, such as staining a tissue using a panel of lectins (often bought as a kit from a commercial source) because they are cheap and readily available; or too ambitious, such as comparing binding patterns of a panel of lectins to a tissue with different diseases. When setting up a research project, we should have a specific and well defined question in mind whose solution is achievable with the time and resources available.

8.2 Mapping lectin binding to normal tissues

Lectin binding has been described for a huge range of normal animal and human tissues (see *Table 8.1*), including basal and other layers of stratified squamous epithelium, neural cells, myelin-producing cells, cells in ganglia, pituitary cells, salivary gland epithelium, myoepithelium, parietal and chief cells of the stomach, pneumocytes, lymphoid follicle centre cells, cells of the bone marrow, macrophages, cortical thymocytes, basal epithelial cells, columnar epithelium of the gastrointestinal tract, mucin cells, endothelial cells, chondrocytes, spermatogenic cells, oocytes, cells of the adrenal medulla, tubules of the kidney, prostatic, ovarian, cervical and endometrial epithelium, and the extracellular matrix of connective tissues. However, there remains a huge need to map the lectin binding and oligosaccharide expression in normal tissues since most investigations have not been systematic ones. If such mapping were to be carried out, lectin binding studies in disease would have a much firmer foundation. Two of the best such papers, concentrating on a single lectin, are those of Cooper (1984) using PNA on human tissues and that of Haines (1993) mapping PNA binding to dog tissues.

Table 8.1: Lectin binding studies to normal human and animal tissues

Tissue	Species	Comments	Reference
Kidney	Human		Silva *et al.* (1993)
	Mouse		Liska (1993)
		Changes with diet	Coppee (1993)
		Changes with ageing	Hanai *et al.* (1994)
	Rat	Newborn vs. adult	Sato (1990)
	Hamster	Normal vs. diabetic	Aguirre (1993)
	Chick embryo	Changes in glycosylation with development	Gheri (1993)
	Quail		Menghi *et al.* (1995)
	Dogfish		Hentschel (1993)
Salivary glands	Rat		Zhang *et al.* (1994)
	Human fetal		Adi *et al.* (1995)
	Dog		Pedini *et al.* (1994)
Gut	Rat/mouse	Changes in Paneth cells	Evans *et al.* (1994)
	Mouse	Changes with development and differentiation	Falk *et al.* (1994, 1995)
	Human	Changes with progression to malignancy	Jass and Roberton (1994)
		Review of mucins coeliac intestinal mucosa	Pittschieler *et al.* (1994)
Nervous system	Human	Changes during development and differentiation	Katz *et al.* (1995)
		Binding to cerebral amyloid and microglia in Alzheimer's disease	Schumacher *et al.* (1994b,c)
	Guinea pig	Cortical neurones distinguished by VVA	Ojima *et al.* (1995)
	Sheep	Microglia	Pennell *et al.* (1994)
	Mammalian	Changes during neurogenesis	Wilson and Wyatt, (1995)
Lung	Sheep endothelium		Abdi *et al.* (1995)
Embryo	*Drosophila*		D'Amico and Jacobs (1995)
Yolk sac	Human		Jones *et al.* (1995)
Prostate	Monkey		Wakui *et al.* (1992)
Storage diseases	Human		DeGasperi *et al.* (1990), Alroy (1994), de Jong (1994)

8.3 Identification of cell lineage

Some lectins bind to a particular type of normal cell with great selectivity, and these can be utilized for fluorescence or magnetic cell sorting, or affinity chromatography to isolate those cells. This is used by immunologists to isolate lymphocyte subpopulations and macrophages, using PNA, HPA, PHA, PWM and ECA.

In solid tissues, identification of the cell type is more difficult. UEA-I commonly binds to human endothelial cells, particularly in capillaries. Some efforts have been made to understand the profiles of normal lectin binding to normal human colorectum (McMahon *et al.*, 1994) and to epithelial cell differentiation and changes following nutritional stress diets on colonic epithelium, using SBA-binding to define structural and functional alterations in glycosylation (Yang, 1994). There is a huge potential for research in this area.

8.4 Mapping lectin binding in disease

8.4.1 Strategy

Rather than simply taking a panel of lectins to look at binding to a series of sections from patients with a particular disease (but too often with different histological type, differentiation, stage in natural history, age of patient, etc., which does not permit any useful conclusion), time should be given to consider the purpose and likely value of the project, other than simply to provide another publication. If diseased tissue is to be examined, then sufficient cases should be selected with comparable clinico-pathological features. It may also be worth consulting the literature and to concentrate on glycosylation types that appear to be of interest in particular tissues: for example, changes in sialylation, expression of $\beta1-6$ branched oligosaccharides, etc., as described in the following sections, appear to be of interest in differentiation of tissues and changes associated with malignancy. A summary of some interesting studies detailing lectin binding to various human cancers is given in *Table 8.2*.

Table 8.2: Lectin binding to a variety of human cancers

Cancer	Lectin	Comments	Reference
Neurological	LCA/PSA	Association with differentiation of astrocytomas	Yang *et al.* (1995)
	PNA	Binding to meningiomas	Marafioti *et al.* (1994)
Cervix	RCA/SBA BSA-I/LTA/UEA-I		Banerjee *et al.* (1995)
	A I A	Association with progression	Remani (1994)
	Various	Binding in CIN and cancer	Griffin and Wells (1994)
Lung	BPA	Different binding to normal, small cell, adeno- and squamous cancers	Sarker *et al.* (1994a)
	DBA	Binding to adeno- and squamous cancers	Fukuoka *et al.* (1984)
	Various	Correlation with metastatic potential	Matsumoto *et al.* (1992)
Prostate	UEA-I	Increased binding in cancer vs. normal	Abel *et al.* (1989), Perlman and Epstein (1990), Nagle *et al.* (1991), Drachenberg and Papadimitriou (1995), Taniguchi (1995)
	HPA	Association with bone metastasis	Shiraishi *et al.* (1992)
Colon	PNA	Marker for TF and T antigen	Campbell *et al.* (1995), Sotozono *et al.* (1994)
	HPA	Association with metastasis	Schumacher *et al.* (1994a), Ikeda *et al.* (1994)
Endothelium	UEA-I	Binding to malignant endothelium	Auerbach *et al.* (1994)
Breast	HPA	Binding associated with metastasis and poor prognosis	Brooks and Leathem (1991), Noguchi *et al.* (1994)
Endometrium	UEA-I	Binding associated with invasion	Ambros and Kurman (1993)
Thyroid	BPA	Binding to normal and cancer	Sarker *et al.* (1994b)

8.4.2 Lectins in distinguishing normal from cancer cells and in identifying cancer cell behaviour

Sialic acids. Sialic acids are found at the non-reducing terminals of oligosaccharide side-chains of most glycoproteins and glycolipids. Their expression, determined by various sialyltransferases, results in cell-specific sialylated glycosylation sequences in normal development (for a review see Roth, 1993). Many studies have demonstrated disturbances in sialylation associated with disease and malignancy (e.g. Sen *et al.*, 1994; Vierbuchen, 1995; Yamashita *et al.*, 1995).

Polylactosamines. Polylactosamines or polylactosaminylglycans are repeating units of *N*-acetyl-lactosamine (galβ1–4 glcNAc β1–3) and their appearance is associated with differentiation and malignancy (Fukuda, 1985; Korczak *et al.*, 1994). They may be detected by LEA, DSA and PWM.

β1–6 branching associated with cancer progression. Progression in rodent and human tumours is commonly associated with changes in glycoprotein glycosylation. In particular, increased β1–6 glcNAc-branching, detected by PHA-L binding, appears to be related to cancer metastasis (Dennis *et al.*, 1987; Fernandes *et al.*, 1991; Korczak *et al.*, 1994).

Tn and Sialyl-Tn. Tn and Sialyl-Tn (STn) are carcinoma-associated carbohydrate determinants expressed on cancer-associated mucins and have the structure galNAc α1-O Ser/Thr and NANA (2-6)α galNAc α1-O Ser/Thr. Expression of Tn and STn has been associated with aggressive biological behaviour and poor prognosis in a variety of cancers, including colorectal, breast, pancreas and gastric cancer (Brooks and Leathem, 1995; Ching *et al.*, 1994; Cho *et al.*, 1994; Miles *et al.*, 1994, 1995; Yang *et al.*, 1994).

HPA binding structures and metastasis. HPA binding has been associated with metastasis and poor prognosis in a number of cancers, including breast (Brooks and Leathem, 1991; Noguchi *et al.*, 1994), colorectal (Ikeda *et al.*, 1994; Schumacher *et al.*, 1994a), stomach (Kakeji *et al.* 1994), mouse melanoma model (Kjonniksen *et al.*, 1994) and oesophageal cancer (Yoshida *et al.*, 1994a,b,c).

8.5 Other approaches

Some ideas for research approaches using lectins to probe glycosylation changes in developing, normal and diseased tissues include the following.

1. Use a single lectin to follow development and differentiation of a particular organ from a single species, at different stages.

   Many of the differentiation and 'onco-fetal' antigens are carbohydrates, or have a major carbohydrate component, such as carcinoembryonic antigen.

   Some normal tissues undergo changes, continuing into adult life, such as the mammary gland which shows dramatic changes during fetal development and pubertal change, menstrual changes, pregnancy and involution.

   Some cancers, such as leukaemias, appear to be a failure of differentiation and the outcome and response to treatment varies with this differentiation. Such functional classifications of disease would be far more helpful than simple morphology and there is enormous scope for new and useful classifications of diseases using lectins.

2. Use a single lectin to map binding to a normal organ from different species (e.g. kidney) or different cell types within an organ.

   This would help to identify tissue- or cell-specific markers that might be essential to the function of a particular cell type.

3. Use a single lectin to map glycosylation in a normal organ under different physiological conditions, such as diet effects on gut, or of cell changes in tissue culture under different conditions.

   Very little is known of the environmental factors that alter glycosylation, presumably through availability of sugars and enzymes within compartments of a cell, or even gene promotion to provide enzymes of glycosylation. Simple experiments are needed, particularly using tissue culture models, to dissect this.

4. Use a panel of lectins with similar nominal sugar specificity, such as gal or galNAc ones, to map sugar expression within an organ.

   The demonstration of differences in lectin binding provides a sensitive first step in tissue analysis, as very minor changes in glycosylation or subpopulations of cells can be detected. Identification of the molecules responsible for the glycosylation differences then requires lectin affinity chromatography of lysed and solubilized tissues. Once the sugar structures and changes are known, it is possible to predict the likely enzyme pathways involved, eventually leading to understanding of the genetic changes.

References

Abdi K, Kobzik L, Li X, Mentzer SJ. (1995) Expression of membrane glycoconjugates on sheep lung endothelium. *Lab. Invest.* **72**, 445–452.

Abel PD, Keane P, Leathem A, Tebbutt S, Williams G. (1989) Change in glycoconjugate for the binding site of the lectin *Ulex europaeus* 1 following malignant transformation of prostatic epithelium. *Br. J. Urol.* **63**, 183–185.

Adi MM, Chisholm DM, Waterhouse JP. (1995) Histochemical study of lectin binding in the human fetal minor salivary glands. *J. Oral Pathol. Med.* **24**, 130–135.

Aguirre JI, Han JS, Itagaki S, Doi K. (1993) Lectin histochemical study on the kidney of normal and streptozotocin-induced diabetic hamsters. *Histol. Histopathol.* **8**, 273–278.

Alroy J, Goyal V, Skutelsky E. (1987) Lectin histochemistry of mammalian endothelium. *Histochemistry* **86**, 603–607.

Alroy J, Castagnaro M, Skutelsky E, Lomakina I. (1994) Lectin histochemistry of infantile lysosomal storage disease associated with osteopetrosis. *Acta Neuropathol. Berlin* **87**, 594–597.

Ambros RA, Kurman RJ. (1993) Association of *Ulex europaeus* agglutinin I binding with invasion in endometrial carcinoma. *Int. J. Gynecol. Pathol.* **12**, 301–306.

Auerbach R, Modzalewski RA, Plendl J, Wang SJ. (1994) From primitive embryonic precursor cells to organ- and tumor-specific vascular endothelial cells: a progress report. *Proc. Ann. Meet. Am. Assoc. Cancer Res.* **35**, 663.

Banerjee S, Robson P, Soutter WP, Foster CS. (1995) Modulated expression of glycoprotein oligosaccharides identifies phenotypic differentiation in squamous carcinomas of the human cervix. *Hum. Pathol.* **26**, 1005–1013.

Brooks SA, Leathem AJC. (1991) Prediction of lymph node involvement in breast cancer by detection of altered glycosylation in the primary tumour. *Lancet* **338**, 71–74.

Brooks SA, Leathem AJC. (1995) Expression of alpha-GalNAc glycoproteins by breast cancers. *Br. J. Cancer* **71**, 1033–1038.

Campbell BJ, Finnie IA, Hounsell EF, Rhodes JM. (1995) Direct demonstration of increased expression of Thomsen-Friedenreich (TF) antigen in colonic adenocarcinoma and ulcerative colitis mucin and its concealment in normal mucin. *J. Clin. Invest.* **95**, 571–576.

Ching CK, Holmes SW, Holmes GKT, Long RG. (1994) Blood-group sialyl-Tn antigen is more specific than Tn as a tumor marker in the pancreas. *Pancreas* **9**, 698–702.

Cho SH, Sahin A, Hortobagyi GN, Hittelman WN, Dhingra K. (1994) Sialyl-Tn antigen expression occurs early during human mammary carcinogenesis and is associated with high nuclear grade and aneuploidy. *Cancer Res.* **54**, 6302–6305.

Cooper H. (1984) Lectins as probes in histochemistry and immunohistochemistry: the peanut (*Arachis hypogaea*) lectin. *Hum. Pathol.* **15**, 904–906.

Coppee I, Gabius HJ, Danguy A. (1993) Histochemical analysis of carbohydrate moieties and sugar-specific acceptors in the kidneys of the laboratory mouse and the golden spiny mouse (*Acomys russatus*). *Histol. Histopathol.* **8**, 673–683.

Damjanov I. (1987) Lectin cytochemistry and histochemistry. *Lab. Invest.* **57**, 5–20.

D'Amico P, Jacobs JR. (1995) Lectin histochemistry of the *Drosophila* embryo. *Tiss. Cell* **27**, 23–30.

DeGasperi R, Alroy J, Richard R, Goyal V, Orgad U, Lee RE, Warren CD. (1990) Glycoprotein storage in Gaucher disease: lectin histochemistry and biochemical studies. *Lab. Invest.* **63**, 385–392.

Dennis JW, Laferte S, Waghorne C, Breitman ML, Kerbel RS. (1987) Beta 1-6 branching of Asn-linked oligosaccharides is directly associated with metastasis. *Science* **236**, 582–585.

Drachenberg CB, Papadimitriou JC. (1995) Aberrant pattern of lectin binding in low and high grade prostatic intraepithelial neoplasia. *Cancer* **75**, 2539–2544.

Evans GS, Chwalinski S, Owen G, Booth C, Singh A, Potten CS. (1994) Expression of pokeweed lectin binding in murine intestinal Paneth cells. *Epithelial Cell Biol.* **3**, 7–15.

Falk P, Roth KA, Gordon JI. (1994) Lectins are sensitive tools for defining the differentiation programs of mouse gut epithelial cell lineages. *Am. J. Physiol.* **266**, G987–G1003.

Falk P, Lorenz RG, Sharon N, Gordon JI. (1995) *Moluccella laevis* lectin, a marker for cellular differentiation programs in mouse gut epithelium. *Am. J. Physiol.* **268**, G553–G567.

Fernandes B, Sagman U, Auger M, Demetrio M, Dennis JW. (1991) Beta 1-6 branched oligosaccharides as a marker of tumor progression in human breast and colon neoplasia *Cancer Res.* **51**, 718–723.

Fukuda M. (1985) Cell surface glycoconjugates as onco-differentiation markers in hematopoietic cells. *Biochim. Biophys. Acta* **780**, 119–150.

Fukuoka K, Yoneda T, Kohnoike Y, Tomoda K, Yoshikawa M, Katada H, Narita N, Imai T, Ioka S. (1994) Lectin binding to non-small cell lung carcinoma. *Lung Cancer Jpn.* **34**, 161–170.

Griffin NR, Wells M. (1994) Characterisation of complex carbohydrates in cervical glandular intraepithelial neoplasia and invasive adenocarcinoma. *Int. J. Gynecol. Pathol.* **13**, 319–329.

Gheri G, Bryk-SG, Sgambati-E, Russo-G. (1993) Chick embryo metanephros: the glycosylation pattern as revealed with lectin conjugates. *Acta Histochem.* **94**, 113–124.

Haines DM. (1993) Peanut agglutinin lectin immunohistochemical staining of normal and neoplastic canine tissues. *Vet. Pathol.* **30**, 333–342.

Hanai T, Usuda N, Morita T, Nagata T. (1994) Light microscopic lectin histochemistry in aging mouse kidney: study of compositional changes in glycoconjugates. *J. Histochem. Cytochem.* **42**, 897–906.

Hentschel H, Walther P. (1993) Heterogenous distribution of glycoconjugates in the kidney of dogfish *Scyliorhinus caniculus* with reference to changes in the glycosylation pattern during ontogenetic development of the nephron. *Anat. Rec.* **235**, 21–32.

Ikeda Y, Mori M, Adachi Y, Matsushima T, Sugimachi K. (1994) Prognostic value of the histochemical expression of *Helix pomatia* agglutinin in advanced colorectal cancer: a univariate and multivariate analysis. *Dis. Colon Rectum* **37**, 181–184.

Jass JR, Roberton AM. (1994) Colorectal mucin histochemistry in health and disease: a critical review. *Pathol. Int.* **44**, 487–504.

Jones CJ , Jauniaux E, Stoddart RW. (1995) Glycans of the early human yolk sac. *Histochem. J.* **27**, 210–221.

de Jong J, van den Berg C, Wijburg H, Willemsen R, van Diggelen O, Schindler D, Hoevenaars F, Wevers R. (1994) Alpha-*N*-acetylgalactosaminidase deficiency with mild clinical manifestations and difficult biochemical diagnosis. *J. Pediatr.* **125**, 385–931.

Kakeji Y, Maehara Y, Tsujitani S, Baba H, Ohno S , Watanabe A, Sugimachi K. (1994) *Helix pomatia* agglutinin binding activity and lymph node metastasis in patients with gastric cancer. *Semin. Surg. Oncol.* **10**, 130–134.

Katz DM, White ME, Hall AK. (1995) Lectin binding distinguishes between neuroendocrine and neuronal derivatives of the sympathoadrenal neural crest. *J. Neurobiol.* **26**, 241–252.

Kjonniksen I, Rye PD, Fodstad O. (1994)*Helix pomatia* agglutinin binding in human tumor cell lines: correlation with pulmonary metastases in nude mice. *Br. J. Cancer* **69**, 1021–1024.

Korczak B, Goss P, Fernandez B, Baker M, Dennis JW. (1994) Branching N-linked oligosaccharides in breast cancer. *Adv. Exp. Med. Biol.* **353**, 95–104.

Liska J, Jakubovsky-J, Ruzickova-M, Surmikova-E, Zaviacic-M. (1993) The use of lectins identified with specific antibodies in lectin histochemistry of NZB/W F1 mouse kidney. *Acta Histochem.* **94**, 185–188.

Marafioti T, Barresi G, Batolo D. (1994) Lectin histochemistry of human meningiomas. *Histol. Histopathol.* **9**, 535–540.

Matsumoto H, Muramatsu H, Muramatsu T, Shimazu H. (1992) Carbohydrate profiles shown by a lectin and a monoclonal antibody correlate with metastatic potential and prognosis of human lung carcinomas. *Cancer* **69**, 2084–2090.

McMahon RF, Panesar MJ, Stoddart RW. (1994) Glycoconjugates of the normal human colorectum: a lectin histochemical study. *Histochem. J.* **26**, 504–518.

Menghi G, Gabrielli MG, Accili-D. (1995) Mosaic lectin labelling in the quail collecting ducts. *Histol. Histopathol.* **10**, 305–312.

Miles DW, Happerfield LC, Smith P, Gillibrand R, Bobrow LG, Gregory WM, Rubens RD. (1994) Expression of sialyl-Tn predicts the effect of adjuvant chemotherapy in node-positive breast cancer. *Br. J. Cancer* **70**, 1272–1275.

Nagle RB, Brawer MK, Kittelson J, Clark-V. (1991) Phenotypic relationships of prostatic intraepithelial neoplasia to invasive prostatic carcinoma. *Am. J. Pathol.* **138**, 119–128.

Noguchi M, Earashi M, Ohnishi I, Kinoshita K, Thomas M, Fusida S, Miyazaki I, Mizukami Y. (1994) nm23 expression versus *Helix pomatia* lectin binding in human breast cancer metastases. *Int. J. Oncol.* **4**, 1353–1358.

Ojima H, Kuroda M, Ohyama J, Kishi K. (1995) Two classes of cortical neurones labelled with *Vicia villosa* lectin in the guinea-pig. *Neuroreport* **6**, 617–620.

Pedini V, Ceccarelli P, Gargiulo AM. (1994) Glycoconjugates in the mandibular salivary gland of adult dogs revealed by lectin histochemistry. *Res. Vet. Sci.* **57**, 353–357.

Pennell NA, Hurley SD, Streit WJ. (1994) Lectin staining of sheep microglia. *Histochemistry* **102**, 483–486.

Perlman EJ, Epstein JI. (1990) Blood group antigen expression in dysplasia and adenocarcinoma of the prostate. *Am. J. Surg. Pathol.* **14**, 810–818.

Pittschieler K, Ladinser B, Petell JK. (1994) Reactivity of gliadin and lectins with celiac intestinal mucosa. *Pediatr. Res.* **36**, 635–641.

Remani P, Pillai KR, Haseenabeevi VM, Ankathil R, Bhattathiri VN, Nair MK, Vijayakumar T. (1994) Lectin cytochemistry in the exfoliative cytology of uterine cervix. *Neoplasma* **41**, 39–42.

Roth J. (1993) Cellular sialoglycoconjugates: a histochemical perspective. *Histochem. J.* **25**, 687–710.

Sarker AB, Koirala TR, Aftabuddin M, Jeon HJ, Murakami I. (1994a) Lectin histochemistry of normal lung and pulmonary carcinoma. *Ind. J. Pathol. Microbiol.* **37**, 29–38.

Sarker AB, Akagi T, Teramoto N, Nose S, Yoshino T, Kondo E. (1994b) *Bauhinia purpurea* (BPA) binding to normal and neoplastic thyroid glands. *Pathol. Res. Pract.* **190**, 1005–1011.

Sato H, Toyoda K, Furukawa F, Ogasawara H, Imazawa T, Imaida K, Takahashi M, Hayashi Y. (1990) Lectin reactivity in the kidney of newborn rat compared to adult rat. *Eisei-Shikenjo-Hokoku* **108**, 78–83.

Schumacher U, Higgs D, Loizidou M, Pickering R, Leathem A, Taylor I. (1994a) *Helix pomatia* agglutinin binding is a useful prognostic indicator in colorectal carcinoma. *Cancer* **74**, 3104–3107.

Schumacher U, Adam E, Kretzschmar H, Pfuller U. (1994b) Binding patterns of mistletoe lectins I, II and III to microglia and Alzheimer plaque glycoproteins in human brains. *Acta Histochem.* **96**, 399–403.

Schumacher U, Kretzschmar H, Pfuller U. (1994c) Staining of cerebral amyloid plaque glycoproteins in patients with Alzheimer's disease with the microglia-specific lectin from mistletoe. *Acta Neuropathol. Berl.* **87**, 422–424.

Sen G, Chowdhury M, Mandal C. (1994) O-acetylated sialic acid as a distinct marker for differentiation between several leukemia erythrocytes. *Mol. Cell Biochem.* **136**, 65–70.

Shiraishi T, Atsumi S, Yatani R. (1992) Comparative study of prostatic carcinoma bone metastasis among Japanese in Japan and Japanese Americans and whites in Hawaii. *Adv. Exp. Med. Biol.* **324**, 7–16.

Silva FG, Nadasdy T, Laszik Z. (1993) Immunohistochemical and lectin dissection of the human nephron in health and disease. *Arch. Pathol. Lab. Med.* **117**, 1233–1239.

Sotozono MA, Okada Y, Tsuji T. (1994) The Thomsen-Friedenreich antigen-related carbohydrate antigens in human gastric intestinal metaplasia and cancer. *J. Histochem. Cytochem.* **42**, 1575–1584.

Spicer SS, Schulte BA. (1992) Diversity of cell glycoconjugates shown histochemically: a perspective. *J. Histochem.Cytochem.* **40**, 1–38.

Strous GJ, Dekker J. (1992) Mucin type glycoproteins. *Crit. Rev. Biochem. Mol. Biol.* **27**, 57–92.

Taniguchi J, Moriyama N, Nagase Y, Kurimoto S, Kawabe K. (1995) Histochemical study of biotinylated lectins in prostatic cancer. *Jpn. J. Urol.* **86**, 1008–1015.

Vierbuchen M, Schroder S, Larena A, Uhlenbruck G, Fischer R. (1994) Native and sialic acid masked Lewis(a) antigen reactivity in medullary thyroid carcinoma. Distinct tumour-associated and prognostic relevant antigens. *Virchows Arch.* **424**, 205–211.

Vierbuchen MJ, Fruechtnicht W, Brackrock S, Krause KT, Zienkiewicz TJ. (1995) Quantitative lectinhistochemical and immunohistochemical studies on the occurrence of alpha(2,3)- and alpha(2,6)-linked sialic acid residues in colorectal carcinomas: relation to clinicopathologic features. *Cancer* **76**, 727–735.

Wakui S, Furusato M, Nomura Y, Asari M, Kano-Y. (1992) Lectin histochemical study of the prostate gland of the rhesus monkey (*Macaca mulatta*). *J. Anat.* **181**, 127–131.

Walker RA. (1989) The use of lectins in histopathology. *Path. Res.Pract.* **185**, 826–835.

Wilson DB, Wyatt DP. (1995) Patterns of lectin binding during mammalian neurogenesis. *J. Anat.* **186**, 209–216.

Yamashita K, Fukushima K, Sakiyama T, Murata F, Kuroki M, Matsuoka Y. (1995) Expression of Sia alpha 2→6Gal beta 1→4GlcNAc residues on sugar chains of glycoproteins including carcinoembryonic antigens in human colon adenocarcinoma: applications of *Trichosanthes japonica* agglutinin I for early diagnosis. *Cancer Res.* **55**, 1675–1679.

Yang K, Fan KH, Newmark H, Lipkin M. (1994) Intermediate endpoints of colonic epithelial cell differentiation following nutritional stress diets. *Proc. Ann. Meet. Am. Assoc. Cancer Res.* **35**, A3701

Yang S, Liu Z, Zhu Y, Chen X. (1995) Labelling and quantitative analysis of six lectin receptors of intracranial gliomas. *Chin. Med. J. Engl.* **108**, 37–43.

Yoshida Y, Okamura T, Shirakusa T. (1993a) An immunohistochemical study of *Helix pomatia* agglutinin binding on carcinomas of the esophagus. *Surg. Gynecol. Obstet.* **177**, 299–302.

Yoshida Y, Okamura T, Yano K, Taga S, Ezaki T. (1994b) Histopathological characteristics associated with long-term survival in stage III esophageal carcinoma. *Cancer J.* **7**, 147–149.

Yoshida Y, Okamura T, Yano K, Ezaki T. (1994c) Silver stained nucleolar organizer region proteins and *Helix pomatia* agglutinin immunostaining in esophageal carcinoma: correlated prognostic factors. *J. Surg. Oncol.* **56**, 116–121.

Zhang XS, Proctor GB, Garrett JR, Schulte BA, Shori DK. (1994) Use of lectin probes on tissues and sympathetic saliva to study the glycoproteins secreted by rat submandibular glands. *J. Histochem. Cytochem.* **42**, 1261–1269.

9 Histochemistry to Localize Endogenous Lectins

9.1 Introduction

The ubiquitous distribution of lectins and their high concentration in some tissues (particularly plant tissues) suggests their importance to the survival and function of cells. In addition, owing to this widespread distribution and high concentration, we might expect their detection in tissues to be easy, but this is not so. The original tool for their localization, described by Monsigny *et al.* (1976), was horseradish peroxidase which binds to mannose-receptors. Since peroxidase is rich in mannose but also is an enzyme it can be visualized directly in a tissue section. This is still a delightfully simple approach to demonstrate mannose-binding lectins. The work of Ashwell and Morell (1974) on clearance of serum proteins according to their glycans, combined with that of Monsigny *et al.* (1976), Simpson *et al.* (1978), Barondes (1978), Kieda *et al.* (1979) and Tanabe *et al.* (1979) led to the development of enzyme, fluorescence and electron microscopy methods to detect cell lectins. These original methods are simple and effective.

Gabius has reported histochemical localization of endogenous lectins in a wide variety of human and animal tissues using his synthesized neoglycoproteins (e.g. Gabius *et al.* 1988, 1992, 1993). However, there are few reports by other groups and this technique applied to light microscopy does seem to give variable results, with weak localized staining or high background, and we have experienced great problems in reproducing meaningful negative controls. Even in tissues that are known to contain high concentrations of lectin, such as plant tissue sections, histochemical localization of neoglycoprotein binding in frozen or in paraffin sections can be very disappointing.

Demonstration of membrane or tissue-bound lectins has been most effective using single cells, such as yeasts (Masy *et al.*, 1992), bacteria (Morioka *et al.*, 1994) and parasites (el-Moudni *et al.*, 1993; Robert *et al.*, 1991; Schottelius, 1992) by fluorescence and particularly using electron microscopy (Lew *et al.*, 1994; Stein *et al.*, 1994; Tanimura and Morioka, 1993). The much greater success achieved in localization of lectins on the surface of single viable cells in contrast with the problems encountered on tissue sections may reflect the fluidity of the cell membrane, enabling endogenous lectins to cluster during incubation and increase the cross-linking and binding of neoglycoproteins. This fluidity can be observed where living cells are incubated with a fluorescent lectin at 4°C to show a fine membrane staining which changes to granularity or capping if they are then incubated at 37°C.

Elucidation of the *in vivo* fate of glycosylated markers has been sought by different functional assays, such as those of Jansen *et al.* (1991), Juneja *et al.* (1992) and Rushfeldt and Smedsrd (1993) and electron microscopy helps to follow the complex events, such as internalization, degradation and recycling.

We rarely know the identity of the natural ligand for lectins, and the form of glycan presented (most suitable sugars, linkage, spatial arrangement, clustering density) on neoglycoproteins is probably most important. Elegant work by Adler *et al.* (1995) to identify the highest affinity target for endogenous lectin on *Entamoeba histolytica*, demonstrated how multivalent neoglycoproteins that bind to one endogenous galNAc lectin do not bind with high affinity to another endogenous galNAc lectin. Polymers of the same monosaccharide may bind 200 000 times more than the monovalent sugar (Adler *et al.*, 1995) but the binding affinity of lectins to these neoglycoproteins is several orders of magnitude less than that between an antibody and its antigen, presumably because the natural ligands for lectins have not yet been identified.

Endogenous lectin binding sites may already be saturated with high affinity ligands and a combined functional–immunohisto-chemical–affinity blotting approach is probably needed for the detection of these lectins. We have described the problems with this approach elsewhere (Schumacher, 1992). A similar situation occurred historically in the histochemical detection of steroid receptors based on ligand binding, which because of its unreliability has now been replaced by immune assays using antibodies against the receptor. To relate the existence of endogenous lectins to biological or clinical behaviour we need reliable lectin assays, probably by immunoassay, and to do this we must first purify endogenous lectins.

Just as we need to isolate and characterize the cellular oligosaccharides recognized by lectins, we need to isolate and characterize the endogenous lectins. Several have been isolated, such as galectins and galaptin, and their detection by immuno-histochemical methods (Akimoto *et al.*, 1995; Ozeki *et al.*, 1995) combined with other techniques (Baum *et al.*, 1995; Inohara and Raz, 1995; Schoeppner *et al.*, 1995) looks very interesting.

9.2 Cellular functions of glycoconjugates

As can be seen from previous chapters, numerous different carbohydrate residues can be detected in tissue sections and Western blots of glycoproteins by the use of lectins. This abundance and diversity of carbohydrate residues should be interpreted in functional terms and, indeed, several functions have been attributed to carbohydrate residues of glycoproteins. They include protection of the protein core from proteolytic enzymes, direction of protein flow within the cell during cellular sorting processes (e.g. mannose-6-phosphate serves as a marker for proteins targetted to the lysosomes), hydration and protection of the cell surfaces by the carbohydrate residues of mucus glycoproteins and proper orientation of membrane glycoproteins within the lipid bilayer. However, probably some of the previous assumptions on the functions of carbohydrate residues, particularly of N-linked glycoproteins, have to be revised since new data indicate that glycosylation can act in a much more specific way than prevously thought (Fiedler and Simons, 1995).

9.3 Endogenous lectins

Because the carbohydrate residues of the cell membrane are located at the extracellular surface of the membrane, one obvious specific function of carbohydrate residues is their interaction with naturally occurring ligands. The lectins found within (animal) tissues are often referred to as endogenous lectins or endolectins. Like any other protein, the presence of endogenous lectins in cell or tissue preparations can be demonstrated in a number of different ways.

9.4 Localization of endolectins by *in situ* hybridization

It is theoretically possible, for example, to demonstrate the presence of endogenous lectins by the technique of *in situ* hybridization. A labelled anti-sense oligonucleotide is required for this approach and (if proper controls are included) the presence of mRNA for a particular lectin can be localized. This approach would be particularly appropriate if the amino acid sequence of an endogenous lectin was known and no antibodies directed against it were available.

So far, this approach has been little explored, but with the rapid advancement of molecular biology it seems likely that it will become more popular. At present, however, sequence data are not available for most lectins, particularly animal and human endogenous lectins. A second limitation of the approach is that localization of mRNA does not necessarily indicate that translation to functional lectin is actually taking place: in most cases it probably is, but care still needs to be taken when interpreting results. For example, the mRNA for a lectin may be present, but no protein is expressed since the synthesis of the lectin has been newly induced and not enough time has passed since induction to allow for its translation.

9.5 Localization of endolectins by conventional immunohistochemistry

Endogenous lectin may be demonstrated very successfully using conventional immunohistochemical techniques. Antibodies (sometimes ready-labelled with fluorescent or enzyme tags or biotinylated) are available against a range of plant and invertebrate lectins, and anti-lectin antibodies may be prepared 'in house' as detailed in Section 2.3 and labelled with biotin, fluorescence or enzymes as detailed in Section 2.4.

No specialist techniques need to be given here. If a suitable antibody is available, any of the techniques listed in Chapter 4 are appropriate to localize anti-lectin antibody binding direct to frozen or paraffin sections of plant, invertebrate or even mammalian tissue. We have, for example, obtained excellent results in localizing production of SNA in different parts of the elderberry plant using this technique. Appropriate controls should, of course, be included. The only limitation to this approach is the availability of a suitable antibody.

9.6 Localization of endolectins through binding of labelled neoglycoproteins

The binding of a labelled, chemically synthesized glycoprotein (often called a neoglycoprotein) to tissue-bound lectin can also be attempted

(Gabius and Bardosi, 1991). The rationale behind this procedure is that endogenous lectin in, for example, a cell or tissue preparation, will bind to the carbohydrate residue of the neoglycoprotein via its carbohydrate binding site. Subsequently, localization of the bound neoglycoprotein can be visualized by a marker (an enzyme label, fluorescent tag or biotin) attached to it using standard histochemical techniques. BSA is usually the carrier protein, and neoglycoproteins are available labelled with biotin or FITC. A practical modification of this technique uses an enzyme carrier to which carbohydrate ligands are covalently linked: so-called neoglycoenzymes (Gabius *et al.,* 1989).

The principle of locating tissue-bound proteins through their active binding site is not a new one. For example, it has been very successfully used for many years in enzyme histochemistry where the distribution of tissue-bound enzyme is indicated by the presence of its coloured reaction product (a simple example would be to incubate a tissue section with DAB–hydrogen peroxide, the commonly used chromogenic substrate for peroxidase: naturally occurring peroxidase in tissue sections would be readily localized by the granular, brown reaction product). This reaction product is due to the catalytic activity of the functional active site of the enzyme, which converts a soluble, colourless substrate into an insoluble, coloured precipitate.

However, binding of a substrate and conversion of the substrate into an end product is a very different concept from binding of a neoglycoprotein to a tissue section. While in enzyme histochemistry, substrate binding triggers the catalytic action of the enzyme and thereby asserts the specificity of substrate binding to a certain degree, binding of a neoglycoprotein to an endolectin in a tissue section does not trigger any visible action. With histochemistry alone, it is therefore not possible to distinguish properly between high and low affinity binding of neoglycoproteins and it is very important to include proper controls for the specificity of any reaction observed: for example to demonstrate competitive inhibition of neoglycoprotein binding to tissue sections by using the monosaccharide used for the synthesis of that neoglycoprotein. In our hands, this is not always successful (Schumacher, 1992) and we have therefore doubted the validity of some results using this technique. Several other reports used the non-biotinylated form of the neoglycoprotein to demonstrate competitive inhibition (see Gabius and Bardosi, 1991, and references therein), however, in our view this is insufficient since non-carbohydrate-mediated binding due to the protein carrier cannot be distinguished from carbohydrate-mediated binding.

Neoglycoprotein histochemistry can be interesting and informative, but we would urge caution in interpreting its results. Neoglycoproteins can be used in any of the standard techniques given in Chapter 4 so no specialist methods need be given here. A number of practical points are, however, listed below.

1. Use of unfixed cryostat sections, cultured cell or even living animals is the best way to identify tissue lectins using neoglycoproteins. Do not use fixed and wax-embedded sections, despite the fact that they have sometimes been used in neoglycoprotein histochemistry, as they present problems.

   During fixation and processing for wax embedding, proteins and probably the active sites of endolectins are denatured. For this reason, enzyme histochemistry rarely works using routinely embedded wax sections.

   The use of living cells or indeed whole animals is probably even more appropriate if cell membrane lectins are to be investigated. Binding of ligands to receptors located at the cell surface often leads to receptor dimerization or receptor cross-linking and subsequent internalization of receptor–ligand complexes. Again, the integrity of these processes can have an influence on the affinity of lectin–glycoconjugate interactions and can most meaningfully be studied in living systems.

2. If the endolectin under study is involved in cell–cell or cell–matrix interaction, as most will be, it is most likely that this endolectin will already be bound to its natural ligand. Any added neoglycoprotein will almost certainly have a lower affinity for the endolectin than its natural ligand and any attempt at localization will be futile. When using cryostat sections, these interactions may be partially overcome by pre-incubation with the appropriate monosaccharide (0.3 M monosaccharide, 30 min incubation, at room temperature). Afterwards, extensive washing against a sugar-free buffer is necessary. However, even with this precautionary step, it cannot be guaranteed that the lectin–ligand complex will dissociate, or indeed will subsequently reassociate with the neoglycoprotein.

3. Like any other protein, endogenous lectins have their optimum pH, ion concentration (Ca^{2+} and Mg^{2+} are probably most critical) and temperature at which they bind best to their ligands. When using cryostat sections, several different buffers (see Schumacher, 1992) with and without ion supplements should be used at a range of temperatures.

4. Always use the appropriate monosaccharide to ascertain the specificity of neoglycoprotein binding.

5. Initially, it might be easier to use FITC-conjugated neoglycoproteins because carbohydrate mediated binding of, for example, a biotin label to receptors in tissues cannot be easily excluded.

References

Adler P, Wood SJ, Lee YC, Lee RT, Petri WA, Schnaar RL. (1995) High affinity binding of the *Entamoeba histolytica* lectin to polyvalent *N*-acetylgalactosaminides. *J. Biol. Chem.* **270**, 5164–5171.

Akimoto Y, Hirabayashi J, Kasai K, Hirano H. (1995) Expression of the endogenous 14-kDa beta-galactoside-binding lectin galectin in normal human skin. *Cell Tiss. Res.* **280**, 1–10.

Ashwell G, Morell AG. (1974) The role of surface carbohydrates in the hepatic recognition and transport of circulating glycoproteins. *Adv. Enzymol. Relat. Areas Mol. Biol.* **41**, 99–128.

Barondes SH. (1978) Developmentally regulated slime mold lectins and specific cell cohesion. *Birth Defects* **14**, 491–496.

Baum LG, Seilhamer JJ, Pang M, Levine WB, Beynon D, Berliner JA. (1995) Synthesis of an endogeneous lectin, galectin-1, by human endothelial cells is up-regulated by endothelial cell activation. *Glycoconj. J.* **12**, 63–68.

el-Moudni B, Philippe M, Monsigny M, Schrevel J. (1993) *N*-acetylglucosamine-binding proteins on *Plasmodium falciparum* merozoite surface. *Glycobiology* **3**, 305–312.

Fiedler K, Simons K. (1995) The role of N-glycans in the secretory pathway. *Cell* **81**, 309–312.

Gabius H-J, Bardosi A. (1991) Neoglycoproteins as tools in glycohistochemistry. *Prog. Histochem. Cytochem.* **22**, 1–66.

Gabius HJ, Bodanowitz S, Schauer A. (1988) Endogenous sugar-binding proteins in human breast tissue and benign and malignant breast lesions. *Cancer* **61**, 1125–1131.

Gabius S, Hellman K-P, Hellman T, Brinck U, Gabius H-J. (1989) Neoglycoenzymes: a versatile tool for lectin detection in solid phase assays and glycohistochemistry. *Analyt. Biochem.* **182**, 447–451.

Gabius HJ, Bahn H, Holzhausen HJ, Knolle J, Stiller D. (1992) Neoglycoprotein binding to normal urothelium and grade-dependent changes in bladder lesions. *Anticancer Res.* **12**, 987–992.

Gabius HJ, *et al.* (1993) Reverse lectin histochemistry: design and application of glycoligands for detection of cell and tissue lectins. *Histol. Histopathol.* **8**, 369–383.

Inohara H, Raz A. (1995) Functional evidence that cell surface galectin-3 mediates homotypic cell adhesion. *Cancer Res.* **55**, 3267–3271.

Jansen RW, Molema G, Ching TL, Oosting R, Harms G, Moolenaar F, Hardonk MJ, Meijer DK. (1991) Hepatic endocytosis of various types of mannose-terminated albumins. What is important, sugar recognition, net charge, or the combination of these features? *J. Biol. Chem.* **266**, 3343–3348.

Juneja HS, Schmalsteig FC, Rajaraman S, Hanson EM, Lee S, Brasher W. (1992) Heterotypic adherence between murine leukemia/lymphoma cells and marrow stromal cells involves a recognition mechanism with galactosyl and mannosyl specificities. *Exp. Hematol.* **20**, 405–411.

Kieda C, Roche AC, Delmotte F, Monsigny M. (1979) Lymphocyte membrane lectins. Direct visualization by the use of fluoresceinyl glycosylated cyochemical markers. *FEBS Lett.* **99**, 329–332.

Lew DB, Songu-Mize E, Pontow SE, Stahl PD, Rattazzi MC. (1994) A mannose receptor mediates mannosyl-rich glycoprotein-induced mitogenesis in bovine airway smooth muscle cells. *J. Clin. Invest.* **94**, 1855–1863.

Masy CL, Henquinet A, Mestdagh MM. (1992) Fluorescence study of lectinlike receptors involved in the flocculation of the yeast *Saccharomyces cerevisiae. Can. J. Microbiol.* **38**, 405–409.

Monsigny M, Roche AC, Kieda C, Midoux P, Obrenovitch A. (1988) Characterization and biological implications of membrane lectins in tumor, lymphoid and myeloid cells. *Biochimie* **70**, 1633–1649.

Monsigny M, Kieda C, Gros D, Schrevel J. (1976) New markers to visualize cell surface glycoconjugates: glycosylated horseradish peroxidase and glycosylated ferritin. *VIth Eur. Congr. Electron Microsc.* **2**, 39–40.

Monsigny M, Roche AC, Midoux P, Mayer R. (1994) Glycoconjugates as carriers for specific delivery of therapeutic drugs and genes. *Adv. Drug. Deliv. Rev.* **14**, 1–24.

Morioka H, Suganuma A, Tachibana M. (1994) Localization of sugar-binding sites in *Staphylococcus aureus* using gold-labeled neoglycoprotein. *J. Histochem. Cytochem.* **42**, 1609–1613.

Ozeki Y, Yokota Y, Kato KH, Titani K, Matsui-T. (1995) Developmental expression of D-galactoside-binding lectin in sea urchin (*Anthocidaris crassispina*) eggs. *Exp. Cell Res.* **216**, 318–324.

Reeves JR, Cooke TG, Fenton-Lee D, McNicol AM, Ozanne BW, Richards RC, Walsh A. (1994) Localisation of EGF receptors in frozen tissue sections by antibody and biotinylated EGF-based techniques. *J. Histochem. Cytochem.* **42**, 307–314.

Robert R, de la Jarrige PL, Mahaza C, Cottin J, Marot-Leblond A, Senet JM. (1991) Specific binding of neoglycoproteins to *Toxoplasma gondii* tachyzoites. *Infect. Immun.* **59**, 4670–4673.

Rushfeldt C, Smedsrd B. (1993) Distribution of colon cancer cells permanently labeled by lectin-mediated endocytosis of a trap label. *Cancer Res.* **53**, 658–662.

Schoeppner HL, Raz A, Ho SB, Bresalier RS. (1995) Expression of an endogenous galactose-binding lectin correlates with neoplastic progression in the colon. *Cancer* **75**, 2818–2826.

Schottelius J. (1992) Neoglycoproteins as tools for the detection of carbohydrate-specific receptors on the cell surface of *Leishmania. Parasitol. Res.* **78**, 309—315.

Schumacher U. (1992) A critical evaluation of neoglycoprotein binding sites *in vivo* and in sections of mouse tissues. *Histochemistry* **97**, 95–99.

Simpson DL, Thorne DR, Loh HH. (1978) Lectins: endogenous carbohydrate-binding proteins from vertebrate tissues: functional role in recognition processes? *Life Sci.* **22**, 727–748.

Stein BA, Shaw TJ, Turner VF, Murphy CR. (1994) Increased lectin binding capacity of trophoblastic cells of late day 5 rat blastocysts. *J. Anat.* **185**, 669–672.

Tanabe T, Pricer WE, Ashwell G. (1979) Subcellular membrane topology and turnover of a rat hepatic binding protein specific for asialoglycoproteins. *J. Biol. Chem.* **254**, 1038–1043.

Tanimura F, Morioka H. (1993) Sugar-binding sites with specificity to *N*-acetyl-D-glucosaminides in middle ear mucosa of the guinea pig. *Eur. Arch. Otorhinolaryngol.* **250**, 337–341.

Appendix A

Recipes

General buffer recipes

Tris-buffered saline (TBS), pH 7.6

60.57 g Tris base [Tris(hydroxymethyl)aminomethane]
87.09 g sodium chloride.

Dissolve in 1 litre distilled water.
Adjust pH to 7.6 using concentrated hydrochloric acid.
Make up to a total volume of 10 litres with distilled water.

Lectin buffer, pH 7.6

60.57 g Tris base
87.0 g sodium chloride
2.03 g magnesium chloride
1.11 g calcium chloride.

Dissolve in 1 litre distilled water, and adjust pH to 7.6 using concentrated hydrochloric acid.
Make up to a total volume of 10 litres with distilled water.

Citrate buffer, pH 6

2.1 g citric acid monohydrate $C_6H_8O_7.H_2O$.

Dissolve in 1 litre distilled water. pH to 6 using 2 N sodium hydroxide solution NaOH.

Phosphate-buffered saline (PBS), pH 7.4

80.0 g sodium chloride
2.0 g potassium chloride
2.0 g potassium dihydrogen phosphate KH_2PO_4
11.5 g disodium hydrogen phosphate Na_2HPO_4.

Dissolve in 1 litre of distilled water and then make up to a total volume of 10 litres.

Sodium acetate buffer

Stock solution A: 0.2 M sodium acetate (anhydrous) in distilled water.
Stock solution B: 0.2 M glacial acetic acid in distilled water.

For a pH 5.2 buffer, mix 80 ml stock solution A + 20 ml stock solution B.
For a pH 4.5 buffer, mix 50 ml stock solution A + 50 ml stock solution B.

Fixative solutions

Formol saline

Mix:
100 ml 40% formaldehyde (filtered)
9.0 g sodium chloride
900 ml tap water.

Neutral-buffered formaldehyde

100 ml 40% formaldehyde (filtered)
900 ml distilled water
4.0 g sodium dihydrogen phosphate monohydrate $NaH_2PO_4.H_2O$
6.5 g disodium hydrogen phosphate (anhydrous) Na_2HPO_4.

Check pH before use (it should be pH 7) and adjust if necessary.

Bouin's fluid

Mix:
750 ml saturated aqueous picric acid
250 ml 40% formaldehyde (filtered)
50 ml glacial acetic acid.
Make up fresh immediately before use – do not store. Do not fix tissues for longer than 24 h.
After fixing, wash tissues in 100% ethanol. (Not suitable for *in situ* hybridization.)

Staining Ouchterlony gels

Coomassie Brilliant Blue stain for Ouchterlony gels

1.25 g Coomassie Brilliant Blue
450 ml 50% methanol in distilled water
50 ml glacial acetic acid.

Dissolve the dye in the methanol / acetic acid / water.
Filter before use.
Store in a tightly capped bottle. Stable at room temperature.

Coomassie Brilliant Blue destaining solution for Ouchterlony gels

Mix:
100 ml glacial acetic acid
100 ml methanol
500 ml distilled water.

Store in a tightly capped bottle. Stable at room temperature.
After use, the discoloured destaining solution can be regenerated by passing through a layer of activated charcoal in a filter paper-lined funnel, and re-used.

EM recipes

EM blocking buffer

This solution blocks potential free aldehyde groups which may act as sites of non-specific binding of the lectin.
Mix:
5% (v/v) normal goat serum
0.1% (w/v) bovine serum albumin
0.1% (w/v) gelatin
in 50 mM glycine in distilled water.

Impregnation solution

Mix:
70 ml ethanol
26 ml distilled water
4 ml glycerol.

Recipes for SDS–PAGE/blotting

All chemicals should be of purest quality. Always use double distilled water or ultrapure water.

Lysis buffer for release of glycoproteins from cells and tissues

60.57 g Tris base
87 g sodium chloride.

Dilute in 1 litre distilled water.
Adjust pH to 7.4 using concentrated hydrochloric acid.
Make up to a total volume of 10 litres with distilled water.
Add 2.61 g phenylmethylsulphonylfluoride (PMSF).

Separating gel buffer, pH 8.8

Dissolve 18.17 g Tris base in 80 ml distilled water.
Then add 4 ml of a 10% (w/v) solution of SDS.
Adjust pH to 8.8 using 12 M hydrochloric acid.
Adjust volume to 100 ml with distilled water then filter.

Store at 4°C. Stable for several months.

Stacking gel buffer, pH 6.8

Dissolve 6.06 g Tris base in 80 ml distilled water.
Then add 4 ml of a 10% (w/v) solution of SDS.
Adjust pH to 6.8 using 12 M hydrochloric acid.
Adjust volume to 100 ml with distilled water then filter.

Store at 4°C. Stable for several months.

Acrylamide/Bis stock solution

Acrylamide is highly neurotoxic. Use gloves and a fume cabinet for weighing out.

Dissolve 30.0 g acrylamide in 50 ml distilled water.
Then add 0.8 g Bis (*N,N*-methylenebisacrylamide).
Adjust volume to 100 ml, then filter.

Store at 4°C. Stable for several months.

Running buffer (10× concentrate)

30.3 g Tris base
144.1 g glycine
Dissolve in 800 ml distilled water, then adjust volume to 1 litre.

pH should be around pH 8.6; if not, check water.
Stable at room temperature for several months.
For use: dilute 10× concentrate 1:10 in water and add 10% SDS to 0.1% (w/v) final concentration.

10% (w/v) SDS stock solution

Dissolve 10 g SDS in 80 ml water (this may require some gentle warming). Adjust volume to 100 ml.

Stable at room temperature for several months.

Reducing sample buffer

1.51 g Tris base
Dissolve in 25 ml distilled water.
Adjust pH to 7.5 using concentrated hydrochloric acid then add:

40 ml of 10% (w/v) SDS solution in distilled water
10 ml mercaptoethanol
0.002 g Bromophenol Blue
10.0 g sucrose or 20 ml glycerol.

Make up to 100 ml with distilled water.
Divide into 1 ml aliquots and store frozen.

Non-reducing sample buffer

Make up exactly as described for reducing sample buffer, but omit the mercaptoethanol.

Coomassie Brilliant Blue staining solution

Mix:
2.5 ml of 1% CBB-R 250 (w/v)
25 ml isopropyl alcohol
10 ml glacial acetic acid
62.5 ml distilled water.

Stable at room temperature for several months.

Destaining solution for Coomassie Brilliant Blue

Mix:
100 ml isopropyl alcohol
100 ml glacial acetic acid
800 ml distilled water.

Transfer buffer for tank Western blotting

Mix:
100 ml running buffer stock solution (10x)
200 ml methanol
800 ml distilled water.

Transfer buffers for semi-dry Western blottting

Anode buffer I.
9.24 g Tris base
Dissolve in 500 ml of 20% (v/v) methanol in distilled water.

Anode buffer II.
1.52 g Tris base
Dissolve in 500 ml of 20% (v/v) methanol in distilled water.

Cathode buffer.
1.52 g Tris base
2.62 g amino caproic acid
Dissolve in 500 ml (v/v) methanol in distilled water.

Blocking buffer for blots.
4% (w/v) bovine serum albumin and 0.05% (v/v) Tween 20 in lectin buffer.

Recipes for IEF

All chemicals should be of the purest quality. Use double distilled water or Ultrapure water.

Acrylamide/Bis stock solution

14.60 g (29.1%) acrylamide
 0.46 g (0.9%) bis.

Dissolve in 50 ml distilled water, filter and store at 4°C for up to 2 weeks.

Anode buffer

0.33 g asparaginic acid
0.37 g glutaminic acid

Dissolve in 100 ml distilled water.
Addition of minute amounts of sodium azide (NaN_3) prevents microbial growth.
Solution stable for 2 weeks at 4°C.

Cathode buffer

0.44 g arginine
0.40 g lysine
0.40 g sodium hydroxide.

Dissolve in 100 ml distilled water (may require gentle warming).
Addition of minute amounts of sodium azide (NaN_3) prevents microbial growth.
Solution stable for 2 weeks at 4°C.

Recipes for staining IEF gels

Staining solution.
Mix:
100 ml of a 1% (w/v) solution of CBB-R 250 in distilled water
400 ml ethanol
100 ml distilled water
100 ml of 20% (v/v) glacial acetic acid in distilled water.

Filter. Store in a tightly capped bottle. Stable at room temperature for several weeks.

Destaining solution.
Mix:
25 ml ethanol
10 ml glacial acetic acid
65 ml distilled water.

Store in a tightly capped bottle. Stable at room temperature for several weeks.

Appendix B

Suppliers

Anderman and Company Ltd (suppliers of Schleicher & Schuell nitrocellulose), 145 London Road, Kingston-upon-Thames, Surrey KT2 6NH, UK.
Tel 0181 5410035. Fax 0181 5410623.

Biometra Ltd, Whatman House, St Leonard's Road, 20/20, Maidstone, Kent ME16 0LS.
Tel 01622 678872. Fax 01622 752774.

Bio-Rad Laboratories Ltd, BioRad House, Maylands Avenue, Hemel Hempstead, Hertfordshire HP2 7TD, UK.
Tel 0800 181134. Fax 01442 259118.

Boehringer Mannheim UK (Diagnostics & Biochemicals Ltd), Bell Lane, Lewes, East Sussex BN7 1LG, UK.
Tel 01273 480444. Fax 01273 480266.

Bradsure Biologicals (UK agents for EY Laboratories), 67a Brooks Street, Shepshed, Loughborough, Leicestershire LE12 9RF, UK.
Tel 01509 650665. Fax 0858 410520.

Dako Ltd, 16 Manor Courtyard, Hughenden Avenue, High Wycombe, Buckinghamshire HP13 5RE, UK.
Tel 01494 452016. Fax 01494 441846.

EY Laboratories Inc., PO Box 1787, San Mateo, CA 94401, USA.
Tel (800) 8210044. Fax (415) 3422648.

Millipore (UK) Ltd, The Boulevard, Blackmoor Lane, Watford, Hertfordshire WD1 8YW, UK.
Tel 01923 816375. Fax 01923 818297.

Pharmacia Biotech, Davy Avenue, Knowlhill, Milton Keynes MK5 8PH, UK.
Tel 0800 318353. Fax 0800 318354.

Raymond A. Lamb, 6 Sunbeam Road, North Acton, London
NW10 6JL, UK.
Tel 0181 9651834. Fax 0181 9614961.

Sigma-Aldrich Company Ltd, Fancy Road, Poole, Dorset BH12 4QH,
UK.
Tel 0800 373731. Fax 0800 378785.

Vector Laboratories Ltd, 16 Wulfric Square, Bretton, Peterborough,
Cambridgeshire PE3 8RF, UK.
Tel 01733 265530. Fax 01733 263048.

Index